НОВЫЕ ПРОЦЕССЫ БУРЕНИЯ ГЛУБОКИХ СКВАЖИН

NOVYE PROTSESSY BURENIYA GLUBOKIKH SKVAZHIN

DEEP-HOLE DRILLING WITH EXPLOSIVES

DEEP-HOLE DRILLING WITH EXPLOSIVES

by
A. P. Ostrovskii

Authorized translation from the Russian

CONSULTANTS BUREAU
NEW YORK

The Russian text was published by Gostoptekhizdat, the State Scientific and Technical Publishing House of the Petroleum
and Mineral-Fuel Industry,
in Moscow in 1960.

ISBN-13: 978-1-4615-8545-9   e-ISBN-13: 978-1-4615-8543-5
DOI: 10.1007/978-1-4615-8543-5

Анатолий Павлович Островский
НОВЫЕ ПРОЦЕССЫ БУРЕНИЯ ГЛУБОКИХ СКВАЖИН

Library of Congress Catalog Card Number 61-17718

CONTENTS

FOREWORD

The development of an effective method of drilling oil and gas wells to depths of 3-5 km and more is a complex problem of great practical interest.

"Deep Hole Drilling With Explosives" is a distinctive and very interesting work of the engineer A. P. Ostrovskii, who more than twenty years ago, by his original suggestions, made a start in investigating the field of so-called non-bit processes of shattering rocks when drilling holes, and, with his co-workers, developed a fundamentally new explosive method of drilling deep holes.

This book discusses the new trend in worldwide application of explosives in technology and the national economy; it presents considerable experimental material on the effect of explosions in solid media, material that is not only interesting to specialists in mining but also to physicists.

Many phenomena discussed by the author and cited in the book still await explanation, but the fact that these phenomena have been already subjected to experimental study undoubtedly adds to our knowledge of one of the most meagerly investigated areas of the science of explosions—concerning the shattering effect of explosions on solid media.

July 29, 1960

Professor M. A. Sadovskii
Corresponding Member of the Academy of
Sciences of the USSR

INTRODUCTION

Together with the perfection of turbine drilling and the introduction of electrical drills in the USSR, new methods have been developed for shattering rock and drilling deep oil and gas wells without the use of a bit. After prolonged work, the first industrially valuable results have been obtained only for one phase of non-bit drilling—for the use of explosives.

Specific peculiarities are inherent in deep drilling.

When drilling oil and gas wells it is necessary to overcome several obstacles arising from various complications: possible slumping of the hole walls, open oil and gas blowouts, the formation of griffons,* catching and tightening of the drilling instrument, and also the inflow of formational water, contaminating the drilling mud. These complications disturb the drilling procedure and delay drilling time or cause accidents, which sometimes lead to the loss of a hole.

The anticipation of many complications during drilling depends on the properties and quality of the drilling fluids, which strengthen the walls of the hole, prevent ejections of gas and oil, and clean the bottom of the hole of broken rock.

The Soviet Union has been first in developing new types of drilling techniques.

The turbine method of drilling holes has accelerated the exploitation of large oil fields in eastern USSR. By means of turbodrills the drilling of multishaft inclined holes has been introduced, permitting the development of oil and gas fields under the floor of the Caspian Sea and under the Volga and Kama rivers.

An electrodrill, proposed by the author conjointly with N. V. Aleksandrov, is used successfully in drilling deep holes. The electrodrill was first manufactured and applied in the Soviet Union. It greatly expands the field of effective use of bottom-drilling motors, especially for drilling holes at depths greater than 3000-4000 m and for holes in complex geologic conditions, where heavy drilling fluids are required.

The design of turbodrills and electrodrills is distinguished by small diameters of the hydraulic turbines and electric motors and great length (more than 50 diameters). Such instruments permit the development of 100 to 400 kw of power with a diameter of 17-25 cm (6-16 kw for each centimeter of diameter).

Soviet scientists developed the theory and design of multistage turbines and first worked out methods of computation and construction of immersible electrical motors for electrodrills, of supplying electrical current to the motors, and a system of automatic control of the drilling procedure under the complex conditions of sinking deep holes filled with liquid.

Drilling by turbodrill and electrodrill in hard rocks is carried on with intense vibrations under loads on the bit up to 30-40 tons, at depths of several thousand meters, at a hydrostatic pressure in the hole approaching 1000 kg/cm^2, and at a temperature near the bottom of a deep hole reaching 150-200°C. Both types of motor work in a medium of rapidly circulating drilling fluid saturated with mud and abraded particles.

---

*Griffons are zones near the shaft of a hole being drilled where the ground is saturated with petroleum gases; they are marked by considerable subsidence of the soil and by the formation of craters.

Immersible hydraulic and electric motors may be said to belong to the group of most overstrained designs of modern energy equipment.

At several oil and gas deposits in the southern part of the USSR, where the rocks are abrasive and plastic, rotary methods of drilling holes have been successfully employed; such methods, as is well known, are most extensively used in the USA. Under certain geologic conditions rotary drilling should be given preference over drilling with bottom motors having high rpm because of the high penetration rate of the drill, the great angular momentum, and several other technological advantages.

At the same time, because of a change from rotary to turbine methods of drilling in the eastern oil fields, where the rocks are hard, such as in Bashkiria and Tataria, the penetration rate of holes 1600-1700 m deep to Devonian oil deposits has been increased five to tenfold. In Tataria holes have been drilled to such depths in 10-15 days, with an average rate of 20 m/hr and a penetration of 40 m without changing the bit; in the Novo-Troitskoe district of the Krasnodar region similar holes have been drilled in 8-12 days, with an average rate of 50 m/hr and a penetration of 200 m without changing the bit.

Economically the most important feature in the development of drilling operations in recent years is the trend toward drilling holes with smaller diameters. This trend has been possible because of the development of turbodrills and then of electrodrills with diameters as little as 170 mm and with bits having diameters no greater than 190 mm; in their energy parameters and their durability, these instruments are in no way inferior to immersible motors and bits of larger diameter.

Despite the progress achieved in drilling holes to depths down to 2000 m, the techniques of drilling deeper holes have remained at a low level for years. At one of the principal eastern fields, the Mukhanovo for example (in the Kuibyshev district), the drilling of a hole to a depth of 2000 m takes a month to a month and a half, but the lower intervals (to a depth of 3100 m) take five to six months more [1]. The expense of drilling in the 2000-3000 m interval in such a hole is more than ten times that of drilling down to 1000 m. On drilling to even greater depths the disparity in expense becomes even more unfavorable, especially in the face of complications that accompany deep drilling.

The drop in effectiveness of drilling with depth is explained also by worsening of the drilling characteristics of most deep-lying rocks because of their compaction through hydrostatic and rock pressure. In order to increase the effectiveness of drilling rock under these conditions, it is necessary to increase the specific bottom power (i.e., the power in proportion to the area of the hole bottom) by achieving the maximum possible ratio of the moment on the shaft of the turbodrill or electrodrill to the rpm of the bit. But the fulfillment of this requirement is attended by the well-known difficulties of transmitting energy to the bottom of a deep hole through immersible hydraulic turbines of small diameter. These difficulties are present in considerably less degree when immersible electric motors are used.

In order to increase the effectiveness of drilling with immersible motors at great depths, it is advisable, under certain geologic conditions, to employ a rock-shattering instrument of some new type. In this regard there is a certain interest in a bit set with large diamonds, such as one tested in France at the Lac field during the drilling of holes from 2500 to 4200 m deep with turbodrills [2]. The results of these tests have shown, under the conditions of the indicated field, the durability of the bits, determined by the amount of hole drilled per bit (from 150 to 500 m), considerable exceeds the durability of rolling cutter bits.

A change in load apparently has little effect on the penetration rate of diamond bits, whereas an increase in load, within certain limits, on rotary bits* is matched by almost a proportional increase in penetration rate. Thus, the combination of on-bottom motors with diamond bits, under proper geologic conditions,** offers promise of increased effectiveness in drilling the lower intervals of deep holes and of drilling any holes of decreasing or small diameter.

Rather simple requirements are demanded of immersible motors when diamond bits are used, since it is possible to compute the characteristics at a higher rpm, three to four times the rpm permissible for a rolling cutter bit.

---

* According to data from the German company "Winter," the rate of rotation of a bit with diamond insets may be raised to 22 m/sec (2200 rpm) with a bit diameter of 190 mm [2].

** In sections composed of nonabrasive or soft rocks or of rocks of medium hardness.

An increase in effectiveness of drilling deep holes may also be attained by using non-bit processes of shattering rocks, using chemical, thermal, electrical, and other sources of energy.

In raising such a question, we start from the universally known inadequacies of modern technology in drilling deep holes with a mechanical instrument of limited endurance.

In the practice of turbodrilling under conditions found in the southern oil and gas fields, as well as in the eastern part of the country, the effectiveness of drilling the lower intervals of holes below 3000 m is considerably diminished, and this leads to the computation of unsatisfactory drilling indices for such holes. For example, at the Karadag gas condensate field (Azerbaidzhan SSR), the drilling of wells to a depth of 2500 m takes 2-2.5 months; drilling from 2500 to 3500 m takes 3-3.5 months, and from 3500 to 4200 m 7-8 months. In such holes the expense of drilling in intervals below 3000 m is 5-10 times that for drilling in the interval down to 2000 m. The average life of the bit, the penetration rate, and the duration of mechanical drilling of such holes in the interval 3000-4200 m are, respectively, 6.76 m, 1.97 m/hr, and 1098 hr, and the average time spent on raising and lowering operations with the drill pipe (for changing worn bits) is 2130 hr.

Frequent and prolonged interruptions in drilling, because of raising and lowering operations, may lead to complications in the shaft of a hole and may require additional expenditure of time to work out the problem and to adjust the drilling mud.

Fundamentally, methods of drilling holes without bits should explore the possibility of uninterrupted shattering of rocks or similar processes, which in their effectiveness may substantially surpass mechanical drilling.

The interest manifested in new methods of drilling is perfectly justified since oil and gas fields differ in complexity and variability of the geologic conditions; it is impossible to state that any one method of drilling, or one instrument, is universally more desirable or economically superior, satisfying all the various requirements of drilling deep holes.

The present book is concerned with the results of investigations and development of an explosive method, proposed by the author for drilling deep holes in sections composed of strong and abrasive rocks. In this explosive method of drilling, the shattering of rocks and the formation of the shaft of the hole are effected by a sequence of explosions of special charges of liquid explosives. The dimensions of the charges used are small in comparison with the diameter of the holes made by underwater explosions. The explosion of each charge shatters a definite volume of rock, and this material is removed from the hole by flushing. During explosive drilling, charges of liquid explosive mixture are prepared automatically immediately next to the bottom of the hole from chemical components that are not explosive in their initial state; these are supplied in stoichiometric proportions.

The methods of placing the charges at the bottom of a hole, which is filled with circulating fluid during drilling, are various. The components of the explosive mixture may be introduced into the hole in special capsules, which are inserted at definite intervals in the drill pipes by the stream of flushing liquid and are exploded at the same intervals on the bottom of the hole.

In another technique charges of explosive mixture without containers are set at the bottom of the hole.

In choosing an explosive method of shattering rocks we took into account the following positive features:

The chemical energy in the explosive charge is transmitted to the bottom of the hole and produces a shock with practically no loss, regardless of the depth of the hole;

The shattering of hard rocks by the direct effect of explosion and explosive disturbance transmitted to rock through a liquid (shock waves, jet streams) is accompanied by the spalling of large rock fragments with a comparatively small expenditure of energy;

During explosive drilling the diameter of the hole being formed may be regulated within broad limits by variation in the distribution of explosive charges on the surface of the hole bottom and by changing the size of charge; this control permits, in particular, the drilling of deep oil and gas wells of the minimum diameter permissible in their exploitation;

Large rock fragments torn from the walls of a hole during underwater explosions may be used for geologic study;

For personal freedom from danger and avoidance of accident, the explosive process of drilling holes may be accomplished by using special chemical components not explosive in their initial state, the use of which is not complicated by specific requirements inherent in an ordinary unitary explosive or in the means of initiating it.

A detonation passing at high speed with very high pressures and temperatures is one of the most complex phenomena employed in modern technology. The pattern of an explosion and, especially, the effect of the explosion on the surrounding medium have been investigated only in comparatively recent time. Important achievements in this field have been accomplished by the Soviet scientists K. K. Andreev, F. A. Baum, A. F. Belyaev, O. E. Vlasov, Ya. B. Zel'dovich, M. A. Lavrent'ev, L. D. Landau, G. I. Pokrovskii, M. A. Sadovskii, N. N. Semenov, K. P. Stanyukovich, Yu. B. Khariton, and others.

The works of the Soviet scientists discuss the fundamental theoretical questions concerning the physics of explosions; these questions are basic for the solution of a number of engineering problems. The scientists have considered the physical factors characterizing specific underwater explosions, the detonation of explosive charges, the expansion of the products of detonation, and the distribution of shock waves in liquid.

The explosive process of drilling takes place under distinctive conditions, differing substantially from the known conditions and methods of using explosives in technology and in war. Therefore, to find a complete solution to the problem of explosive drilling in the results already obtained in the scientific study of explosions has been impossible; it has been necessary to study and make experimental tests on the shattering of rocks and the formation of holes by underwater explosions.

The author has attempted to embrace a wide range of questions to give a qualitative view of the new processes of shattering rocks and of the problem of drilling deep holes by explosions.

Chapter 1 is devoted to a survey of some processes of shattering rocks by impulses of high pressures of various intensities and durations and of thermal processes of shattering rocks under conditions approximately the same as those in a deep hole filled with liquid.

The phenomena appearing in rocks that have been subjected to explosive charges, transmitted directly and through a liquid medium, and also experiments of forming holes by explosions on models and under natural conditions of rocks in place, the basic parameters of underwater explosions, the zones of influence of such explosions, and the procedure of forming holes by explosions are considered in Chapters 2 and 3.

Chapters 4 and 5 have to do with factors affecting the operation of explosive charges and the transmission of detonations between charges under hydrostatic pressure in a space limited by the walls and bottom of a deep hole.

Erosion on the bottom and movement of large particles of rock in the hole, particles derived from the explosion (and from other pulsing tensions in the rock), are discussed in Chapter 6.

The technical processes of explosive drilling are described in Chapter 7.

Chapters 8 and 9 are devoted to questions on the technology of the explosive method of drilling deep holes, on the results of experimental drilling, and on the evaluation of the potentials of the method.

The explosive process of drilling was first proposed and carried out in the Soviet Union. In the course of its development and testing extensive sections of deep holes were drilled with no assistance from any mechanical rock-breaking instrument; the results, under certain circumstances, were superior to work done with a drill bit. A substantially new trend in explosive drilling has been established. The effectiveness of the process is still inadequate and does not correspond to the potential possible in explosions. This fact is explained by the meagerness of studies made on the complex phenomena of underwater explosions and on the effect of such explosions on the surrounding medium in a deep hole under high hydrostatic pressure.

The development of the explosive method of drilling deep holes has been carried on under the direction of the author by the combined designers and scientific personnel of the All-Union Scientific-Research Institute of Drilling Techniques.

In developing the explosive method of drilling holes V. D. Kruglov, A. I. Gol'binder, Ya. I. Shnapir, V. L. Shukhman, G. V. Vladimirov, A. A. Ratov, V. V. Grachev, S. E. Rashkov, A. V. Khlebnikov, N. S. Polukhin, and others worked on and tested explosive capsules and charging-feeding apparatus; they also took part in setting up the technical procedure for drilling a hole by explosives.

In explaining the effect of disseminated explosive charges (on the floor of the hole) on the effectiveness of the explosions and in seeking a basis for explosive drilling of holes with small diameter, the author made use of data obtained by E. B. Kagan.

The author express his sincere gratitude to all the co-workers listed.

Great aid was obtained from the petroleum production organizations of Bashkiria and the Kuibyshev Oblast in performing the experimental work. To the collective personnel of these organizations, in the persons of S. I. Kuvykin and V. I. Muravlenko, the author proffers his thanks.

The possibility of drilling deep holes by explosions was foreseen by Academician N. N. Semenov, who supported this technical idea during its conception. N. N. Semenov and Professor M. A. Savodskii, corresponding member of the Academy of Sciences of the USSR, made a number of valuable suggestions in connection with the experimental investigations of several phenomena during underwater explosions.

A SURVEY OF SOME PROCESSES OF SHATTERING ROCKS

The present chapter has to do with processes of shattering rocks, as proposed by the author and other workers, by means of impulses of high pressure of various intensities and durations, characterized by short-period application and removal of stress; it also considers thermal processes of shattering rocks.

A. Shattering Rocks by Impulses of High Pressure

An impulse stress differs from an ordinary stress by the suddenness of application and by the short duration of its action.

Here we consider impulse stresses of rather high intensity—sufficiently high to produce shattering and irreversible changes in a solid body, such as rock.

Impulse Jets of Liquid. The first attempt to develop a method of sinking holes without a bit by shattering rocks with high-pressure jets of liquid was made by the author in 1939. Experiments showed that, in order to secure effective erosion and spalling of strong rocks, the pressure needed to force liquid in a stream from a nozzle must be several thousand kilograms per square centimeter. Under these circumstances, even when the nozzle is of small diameter, such as 2 mm, and the pressure is near 4000 kg/cm^2, the required power (without considering loss) exceeds 1500 horsepower. A decrease in power was found possible by changing from continuous action of the stream on the rock to impulse streams, which act for short periods separated by intervals of considerably greater length.

The mechanical effect of impulses of individual high-pressure jets and also the behavior of jets in air and in a liquid environment were studied, for various diameters and configurations of nozzles, on blocks of different rocks by means of a "ballistic" stand and a stand arrangement with a mechanical pressure multiplier.

In the combustion chamber of the ballistic stand (Fig. 1) a pressure of 3000 to 8000 kg/cm^2 was created by the combustion of rapidly burning powder charges or of special spontaneously igniting hot liquid mixtures over a surface of ejecting liquid; the combustion led to the evolution of large quantities of heat and gas.

The impulse of pressure imparted by a jet of liquid (a volume of liquid forced through a nozzle 2 mm in diameter—about 100 ml) was sufficient to make a hole in a metallic bar several millimeters thick. Oscillograms of pressure in the combustion chamber and in the liquid jet at its encounter with the bar are shown in Fig. 2. These data were used to make the jet arrangement, which was designed to operate at a frequency of several tens of cycles per minute. The combustion chamber for this arrangement is shown diagrammatically in Fig. 3. The individual introduction of the liquid oxidizing agent, fuel, and water into the chamber was accomplished by means of high-pressure piston dosage devices and of a pump.

For the study of the mechanical effect of high-pressure jets, use was also made of a stand with a mechanical pressure multiplier, which consisted of a hydraulic pump with pneumatic or hydraulic drive. The multiplier permits one to increase the pressure of the liquid supplied to the drive of the pump; this pressure is proportional to the ratio of the cross-sectional areas of a large and a small piston.

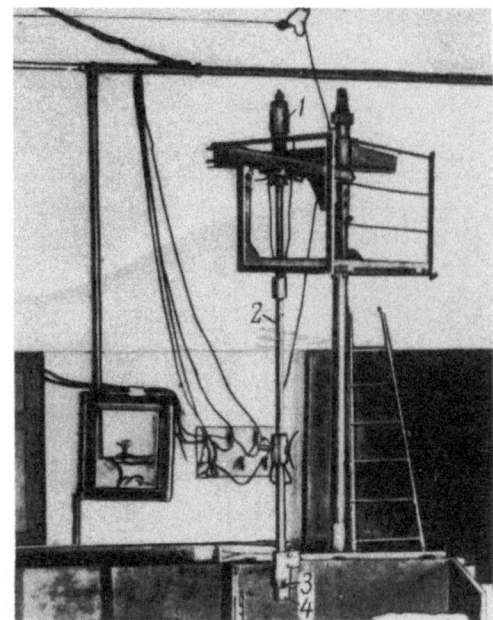

Fig. 1. Ballistic stand. 1) Combustion chamber; 2) shaft of stand; 3) pressure transmitter; 4) nozzle.

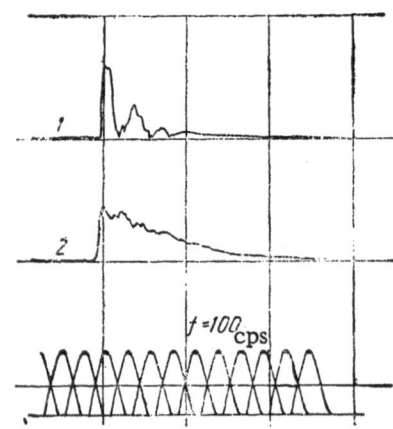

Fig. 2. Oscillograms of pressure. 1) In the combustion chamber of the ballistic stand (P_{max} = 6350 kg/cm^2); 2) in the high-pressure liquid jet where it encountered the bar (P_{max} = 3420 kg/cm^2).

In contrast to the ballistic stand and the jet arrangement, the mechanical pressure multiplier ejected liquid (in volumes up to 100 ml) with a constant pressure of 5000 kg/cm^2 at two cycles per minute and with a pressure of 1000 kg/cm^2 at a frequency of 75 cycles per minute.

In order to obtain high-pressure jets of liquid at the bottom of a hole, it was proposed that the energy of special liquid fuels be used, the material being supplied to a combustion chamber in an on-bottom apparatus [3] according to the principles of the ballistic stand and the jet arrangement (Figs. 1 and 3), or that a mechanical pressure multiplier be used, also on the bottom of the hole [4].

The mixture consisted of liquid oxidizing agent and fuel, taken in the proportion securing the necessary behavior during burning; ignition was effected by introducing a liquid chemical igniter in the fuel or in the chamber (at the site where the mixture was formed).

For a sequential treatment of a given section of hole bottom with impulse jets of liquid by means of a pressure multiplier or by a jet device of the ballistic type, a suitable disposition of nozzles and a rotation of the on-bottom apparatus were required.

The mechanical effect of high-pressure jets flowing at velocities up to 1000 m/sec and at pressures up to 5000 kg/cm^2 was examined on limestone and dolomite (from quarries in the Oka region), on sandstone, on granite (Yantsevskoe deposit), and also on clay (with disturbed natural structure). These experiments established the fact that the nature of the shattering of rocks by high-pressure jets depends on the physical-mechanical properties of the rock, the mineral composition, and the structural and textural features. Strong rocks in which the grains are of the same mineral as the cement binding the grains are shattered with the consequent formation of relatively large fragments when acted on by jets of liquid. Such rocks, spalling off, form conical depressions with an apex angle considerably greater than 90°. Limestone and dolomite belong to this type of rock (Fig. 5).

In strong rocks of inhomogeneous structure, in which the cementing material is considerable softer than the grains of other minerals, the effect of a jet differs from that described. In this case the rocks are strongly shattered with no noticeable spalling, and a cylindrical opening is formed, with a diameter equal to four or five times the diameter of the jet and having a well-defined mouth. Such rocks include sandstones (Fig. 6) consisting of particles (grains of quartz) with high sclerometric hardness and of soft cement (argillaceous, calcareous, and similar material).

Granite belongs to the intermediate group, in which the difference in physical-mechanical properties between grains and cementing material is not great (Fig. 7). For some values of pressure impact this rock behaves like a sandstone; for higher pressure values it responds like a limestone.

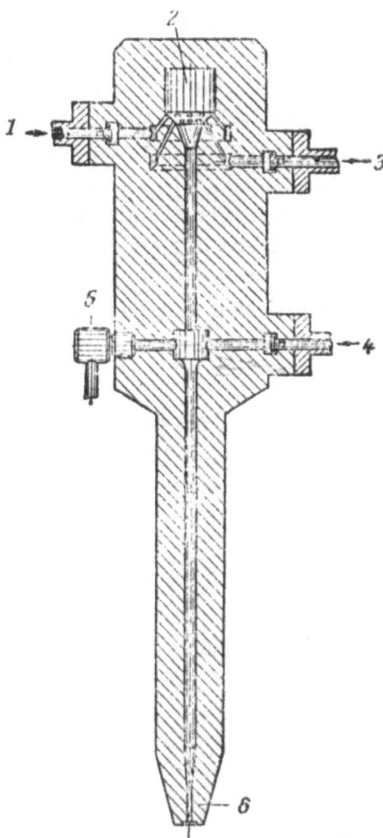

Fig. 3. Combustion chamber of the jet device. 1) Entrance for oxidizing agent; 2) combustion chamber; 3) entrance for fuel; 4) supply of water; 5) pressure transmitter; 6) nozzle.

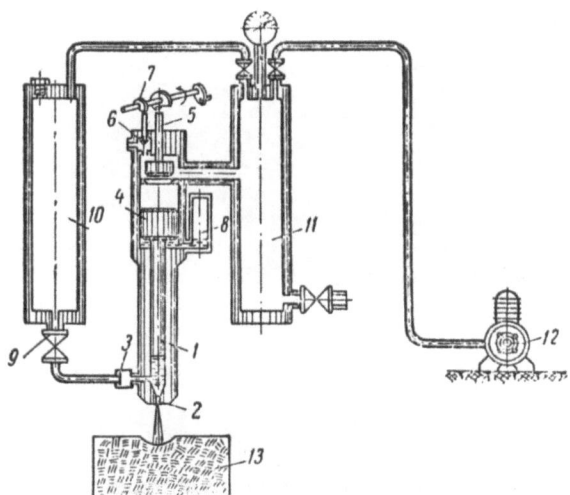

Fig. 4. Stand with mechanical pressure multiplier. 1) High-pressure piston; 2) nozzle; 3) check valve; 4) low-pressure piston; 5) main valve; 6) escape valve; 7) cam shaft; 8) hydraulic shock absorber; 9) globe valve; 10) lubricator; 11) receiver; 12) compresson; 13) block of rock.

In plastic rocks (clay, gypsum, anhydride, siltstone) the effect of the jets is similar to that observed in sandstones, but differs considerably in the zone of scouring. Attempts to form model holes in comparatively strong blocks of hard rock (1.0 × 1.5 × 1.0 m) led to shattering of the rock after a few impulses of high-pressure jets of liquid (Fig. 8).

Some of the relations characterizing the effectiveness of high-pressure impulses of liquid jets on granite (temporary resistance to uniaxial static compression near 2000 kg/cm^2), moving through air and water at a maximum pressure of the jet device in the combustion chamber of 3500 kg/cm^2, are shown in Figs. 9 and 10.

Of recent work we may point out the investigations of the properties of nonimmersed jets carried out at the Institute of Mining of the Academy of Sciences of the USSR, under the direction of A. N. Zelenin. These studies support the view that it is possible to cut rocks with jets of water flowing under high pressure at supersonic velocity. Hydraulic compressers were used in the experiments to obtain jets of water of high velocity.

Experiments have established the optimum diameter of the jets (0.8-1.0 mm) and the distance between nozzle and rock (not greater than 4-5 cm). The width of the cut made by a jet is 3-5 mm. The depth of a cut in strong rock, such as granite, increases with an increase of pressure according to a linear function. At a pressure of 2000 kg/cm^2 the depth of cutting for a single penetration, at a feeding rate of 1.4 cm/sec, amounted to 30 mm for granite (crushing strength of 1300-1600 kg/cm^2), 74 mm for marble (crushing strength of 800 kg per cm^2), and 97 mm for limestone (crushing strength of 600 kg/cm^2).

It should be noted that, according to data from the same source, individual rocks (such as some types of shales), at pressures up to 2000 kg/cm^2, are not susceptible to cutting by jets of water.

Impulses of Pressure in Liquid by Collapse of Vacuum Cavities. To shatter rocks it is possible to use the hydrostatic pressure of the liquid filling the hole (to some depth). For this purpose special hermetically sealed capsules, from which the air has been first removed to the desired state of evacuation, may be placed in the hole.* These capsules are broken by falling against the bottom of the hole at a certain

*Proposed conjointly by the author and E. B. Kagan.

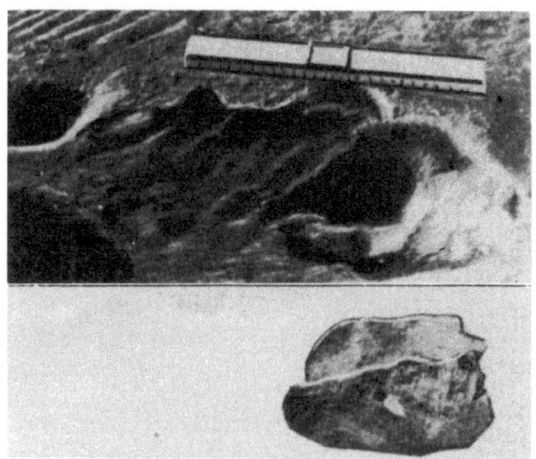

Fig. 5. Spall of limestone. One may see traces of the impact of liquid jets.

velocity or by other means. When the capsule breaks, there occurs an intense implosion into the vacuum cavity, which is immediately next to the rock. The liquid about the cavity acquires a great velocity and the rock is shattered by impulses of very high pressure that are thus communicated to it. This phenomenon, which has been studied in considerable detail, is encountered during cavitation in liquids near the surface of a solid body, when the development and collapsing of empty bubbles (or filled with vapors of the liquid) cause shattering at the indicated surface; this process is called cavitation (or cavitational erosion).

The theoretical value of the pressure P in kilograms per square centimeter during the collapse of a spherical vacuum bubble in an ideal incompressible liquid is given by the formula of Rayleigh [5]:

$$P \approx 0.163 \left(\frac{R_0}{R}\right)^3 P_0,$$

where R_0 is the initial radius of the vacuum sphere, R the radius at any particular moment (during collapse of the sphere), and P_0 the hydrostatic pressure.

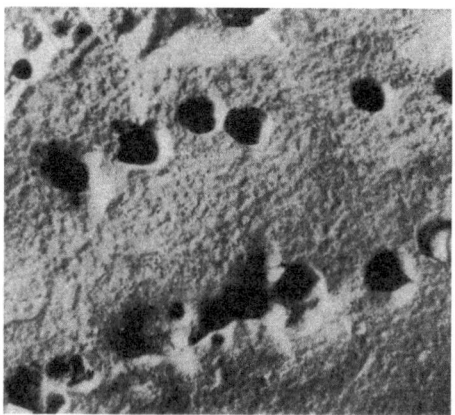

Fig. 6. Scouring of openings in sandstone.

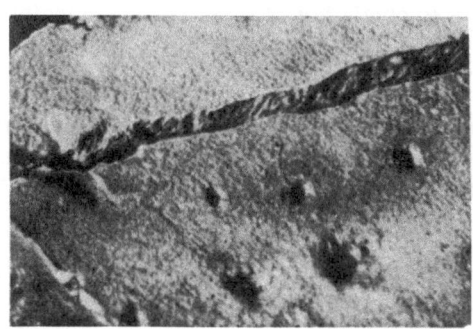

Fig. 7. Scouring of openings in granite and a crack in the block along the openings.

Fig. 8. Shattering of a block of limestone by impulses of liquid jets.

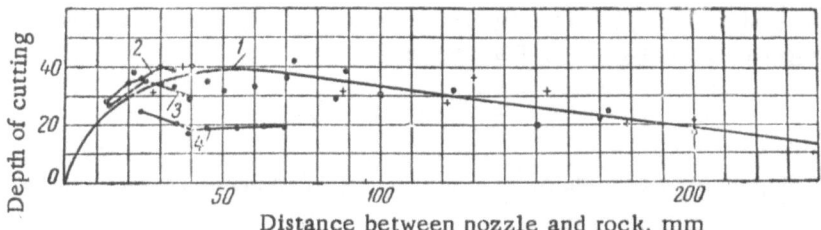

Fig. 9. Relation of depth of cutting in granite to the distance between nozzle and rock. 1) With the liquid (water) jets traveling in air and through a layer of water (diameter of nozzle, 3.15 mm); 2) with the liquid jets moving through a layer of water (diameter of nozzle, 3.10 mm); 3) same as last with a nozzle diameter of 2.25 mm; 4) same with a nozzle diameter of 5.23 mm. ●) Movement of jets through a layer of water; +) movement of jets in air.

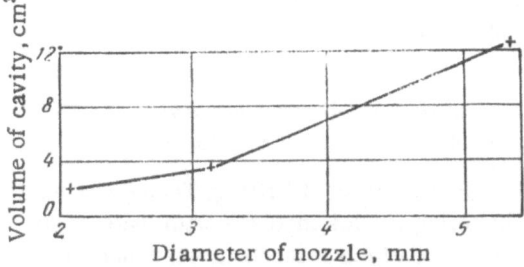

Fig. 10. Relation of volume of excavated cavity in granite to diameter of nozzle (movement of liquid jets through a layer of water).

From the formula of Rayleigh it follows that a considerable pressure develops a liquid from a spherical vacuum bubble with an initial radius of, say, 50 mm and collapsing in a medium with a hydrostatic pressure of 150 kg per cm²:

Radius R, of the contracting sphere, in mm	10	5	3
Pressure, P, in liquid during collapse of vacuum sphere, in kg/cm²	3000	24,000	110,000

It is necessary to bear in mind that the effect of an impulse of pressure on the surface of a solid body depends on the distance to the geometric center of the vacuum bubble. The movement of the liquid mass during collapse of a bubble of constant form is directed toward the center, and the pressure quickly drops in proportion to its distance from the boundary of the bubble. In addition, because of Archimedean forces, the bubble tends to move away from the solid surface (bottom of the hole). Therefore, when conditions are unfavorable, the zone of highest pressure may be found at some considerable distance from the bottom of the hole.

In the Rayleigh formula no attention is paid to the compressibility of the liquid. Consideration of the compressibility also leads to a lower computed value of pressure during collapse of the bubble. One should also anticipate the possibility of a secondary effect, the impact from droplets of liquid against the exposed surface of a solid body. Should such impact occur, an extremely high pressure is developed at the point of collision; the value of this pressure was found by Rayleigh with due consideration to the compressibility of the liquid. The local pressure during such a collision is

$$P' = v\sqrt{\varrho E}$$

where v is the velocity of the droplets of liquid (at the termination of the collapsing process in the bubble this velocity is very large), and ρ and E are, respectively, the density and the bulk modulus of the liquid.

In order to determine approximately the shattering effect from the collapse of a spherical vacuum cavity, one may compute the energy given off in this process and compare it with the shattering effect of an explosive charge of the same energy. In the collapse of a cavity, energy A is given off, equal to the product of the hydrostatic pressure and the volume of the cavity:

$$A = \frac{4}{3}\pi R_0^3 P_0.$$

Using the same values for P_0 and R_0 (150 kg/cm² and 50 mm), we obtain a value of about 750 kg-m for A.

The energy of 1 kg of TNT is approximately equivalent to $4 \cdot 10^5$ kg-m. Thus, the store of energy in a vacuum cavity corresponds to a charge of TNT as shown:

$$q = \frac{750}{4 \cdot 10^5} 1000 \approx 2\,\text{g}.$$

Figure 11 shows diagrammatically the possible distribution of vacuum "charges" on the surface of the hole bottom.

As experiments with equivalent charges of explosives have shown, the central part of the bottom surface (the core) shatters because of brisance and explosive disturbances (shock waves, currents) transmitted by the liquid during explosion of the charge and distributed over a zone of specific radius.

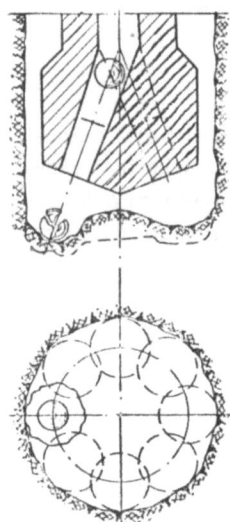

Fig. 11. Distribution of vacuoles on the bottom surface of a hole.

A fundamental feature of the investigated process of shattering rocks is that, for any given vacuum cavities, the store of energy and the impulse of pressure in the liquid during collapse of the cavity increase with increase in hydrostatic pressure, i.e., with depth of the hole.

Ultrasonics.* The attention of investigators has been attracted by the idea of using acoustical waves in ultrasonic, sonic, and infrasonic ranges for shattering rocks when drilling holes.

The use of sonic and infrasonic vibrations during vibratory-percussion of vibratory-rotary drilling increases the penetration of the bit to a certain extent.

Work at the All-Union Scientific-Research Institute of Drilling Techniques has established the fact that the intensity of shattering in rock increases with increase in frequency of vibration, but vibrators, the action of which is mechanical, cannot develop a frequency of more than 80-100 vibrations per second.

Vibrations of an operating instrument with frequencies of 10,000-20,000 vibrations and more per second may be produced, in principal, by ultrasonic methods.

The theoretical use of ultrasonics is not limited to vibratory or vibratory-rotary instruments acting on rocks, since acoustical fluctuations of certain frequencies and intensities in a liquid give rise to cavitational effects and, in this way, also shatter rocks. The shattering of solid bodies such as rocks by the propagation of ultrasonic waves, without the effect of cavitation, has not yet been studied adequately.

Ultrasonic waves are used for cutting and making small manufactured articles from hard materials (diamond, hard alloys, tempered steel, etc.), but the breakdown of the material is not due to ultrasonic propagated through the material, but is rather the result of ultrasonic cavitation. The rate of destroying solid bodies by this process is very low, being measured in millimeters and parts of a millimeter per minute.

Since the crushing strength of rocks is considerably greater than the tensile strength, it may be assumed that shattering in the rock will occur at the passage of the negative half of the waves (rarefaction waves), but to accomplish shattering of rock by direct ultrasonic action, the intensity of a wave must be on the order of tens of thousands watts per square centimeter ($1\,\text{w/cm}^2 = 1 \cdot 10^7\,\text{erg/sec} \cdot \text{cm}^2$). Furthermore, the existing means of generating ultrasonic waves (without concentration) permit the development of but one-thousandth this intensity.

In regard to the shattering of rock by ultrasonic cavitation, there are not sufficient experimental data to make even the most approximate evaluation of the effectiveness of such a process. However, the conditions under which cavitation develops and the physical nature of cavitational breakdown of material have been studied in considerable detail.

When intense ultrasonic vibrations pass through a liquid, compression and rarefaction occur in the liquid, such as produced by the fluctuations of a vibrator, changing their dimensions as they advance. A drop in pressure

---

*Based chiefly on information from the works of V. O. Mal'chenok and O. M. Sumarokov [6], and also of A. P. Pinsker, D. E. Takamlik, and M. K. Kogan [7].

to the critical value, corresponding at a given temperature to vapor formation in the liquid, is accompanied by the growth of tensional stresses, capable of rupturing the liquid and producing numerous cavities (bubbles). When rarefaction gives way to compression, the bubbles that have formed collapse with great force, producing cavitational deterioration in the material. The collapse of each cavitational bubble at the boundary between the liquid and the solid body produces a hydraulic impact in the microscopic zone at the surface of the cavity. Repeated impacts lead to disintegration of the solid body at increasing depths inside the body.

According to Rayleigh, the pressure arising adjacent to bubbles reaches a maximum at a distance of $1.58R$ from the center of the collapsed cavity. The theory of Rayleigh points to the possible development of very high pressures in the vicinity of collapsing bubbles, reaching, for example, values of $40,000$ kg/cm^2 at an external pressure of 2 kg/cm^2 and a ratio of $R_0/R = 50$. In this process the cavity collapses at the rate of 4000 m/sec, and the time for collapsing, with $R_0 = 0.1-1.0$ mm, ranges from 10^{-5} to 10^{-4} sec.

In general, the theory and mechanism of shattering solid bodies are the same in this instance as that described above for the collapse of vacuum cavities under hydrostatic pressure in a hole.

The process of cavitational disintegration of rocks follows a definite sequence: first there occurs disintegration of the softer constituents; then microfractures develop between individual grains and crystals; finally fragments spall off. The products of disintegration form colloidal mixtures of finely dispersed mineral particles in the liquid.

From works devoted to the study of erosion in the cavitational zone [6, 7], it is seen that all materials are subject to erosion, including rocks. For harder materials the rate of erosion is somewhat less.

The development and intensity of the cavitational process depend on many factors; for a given liquid the growth of cavitational pockets is possible only within a narrow range of external pressure and pressure within the bubbles. At pressures of $5-7$ kg/cm^2 cavitational action practically dies out in water.

According to L. Bergman [8], who refers to a great number of special papers, erosion and dispersion of solid bodies during the propagation of ultrasonic waves through a liquid cease at high pressures and in a vacuum.

Thus, we may conclude that attempts to drill deep liquid-filled holes by means of ultrasonic (or sonic) cavitational disintegration of the rocks are not on firm footing, and the direct disintegration of rocks by acoustical waves is impracticable until special small-scale generators are built, which can be immersed in the hole and can generate ultrasonic waves with intensities measured in thousands of watts per square centimeter.

Electrical Discharge in Liquids. The possibility of using hydrodynamic shocks arising during electrical discharge in a liquid for disintegrating rocks has been pointed out by L. A. Yutkin [9].

The fact that such shocks develop has been known for a long time. Even as early as the second half of the eighteenth century the English investigator Priestley observed the ejection of liquid from a vessel during electrical discharge. Since then many scientists have attempted to use this phenomenon for some practical purpose.

Investigations of hydrodynamic shock processes due to electrical discharges in liquids have been carried on by the Petroleum Institute of the Academy of Sciences of the USSR [10], and by the All-Union Scientific-Research Institute for Methods and Techniques of Prospecting [11]. Prior to these investigations the peculiarities of electrical discharges in liquids and the effect of such discharges on the surrounding medium had been but imperfectly studied; the literature discussed chiefly questions referring to the theory and experimental investigations of developing a spark-discharge channel in gases.

Most investigators believe that, in explaining electrical disruption of a liquid, the streamer theory of Leb may be applied [12], a theory based on examinations of disruption of gas at atmospheric pressure.

In amending this theory, P. A. Kulle and P. V. Ponomarev [11] have determined the physical essence of the electrohydrodynamic effect: at the instant a discharge occurs because of elastic collision of individual ions with neutral molecules, the energy of the electrical field is changed to kinetic energy in the molecules and the pressure rises in the discharge channel. Exchange phenomena also foster increase of pressure about a spark; this exchange comes about because fast-moving ions give up their charge on collision with the molecules and are converted to neutral molecules, still maintaining their kinetic energy [13].

On the basis of theoretical analysis and experimental investigations carried out at the All-Union Scientific-Research Institute for Methods and Techniques of Prospecting, the authors have concluded that, for disintegrating a solid body, the most economical procedure is found in using comparatively short impulses, not exceeding a few microseconds in length. Impacts of longer duration lead to dissemination of the energy and to a decrease in mechanical efficiency of the disintegration process.

N. I. Titkov and his co-workers [10] discovered that the disintegration effect of an electrical discharge in liquid is due to mechanical processes associated with the development and propagation of shock waves in the surrounding medium and to the action of these waves on a solid body. These investigations, and also the works of N. N. Dolgopolov (All-Union Scientific-Research Institute of Drilling Techniques), have shown that a discharge occurring within a very short interval of time (10^{-6}-10^{-7} sec) has a mechanical effect on a solid medium similar

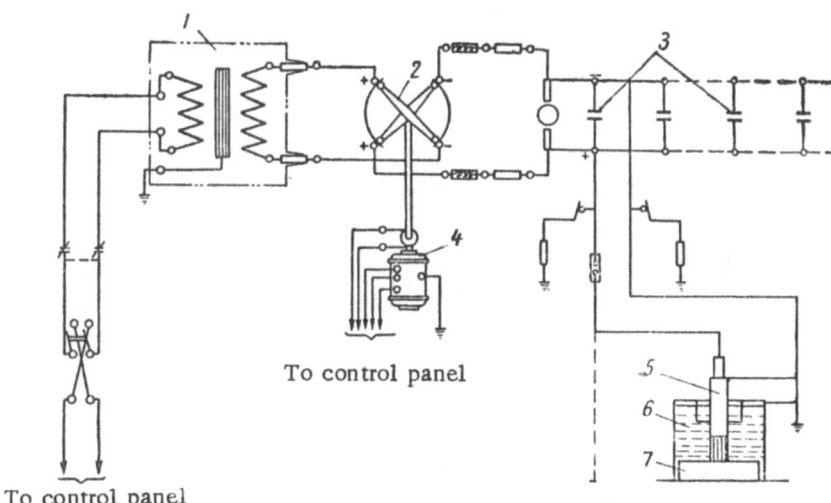

To control panel

To control panel

Fig. 12. Diagram of an impulse generator. 1) High-voltage transformer 2) mechanical rectifier; 3) condensers; 4) motor; 5) drill; 6) reservoir; 7) test material.

to that of an explosion. The parameters of a shock wave forming during discharge depend on the amount of energy evolved, the duration of the discharge, and the properties of the surrounding medium. For example, the energy evolved during the explosion of one gram of TNT (about $0.5 \cdot 10^3$ kg-m) may be obtained from an electrical discharge where the potential across the condensers of the discharge circuit is 100 kv and the capacitance of the condensers is 1 μf. From this we find that the value of instantaneous power is

$$N = \frac{0.5 \cdot 10^3}{10^{-6} \cdot 10^2} = 0.5 \cdot 10^7 \text{ kw},$$

where 10^{-6} sec is the time of discharge in the liquid, according to data of Frungel [14].

At some value of pressure at the front of a shock wave traveling through a liquid, and at the impact of this front against a solid body in the liquid, tensional stresses exceeding the tensile strength of the material may arise. In addition, disintegration due to fatigue of strong material may be produced by repeated discharges.

Preliminary experimental investigations on the disintegration of samples of various rocks by crushing them were made with an impulse generator, a diagram of which is shown in Fig. 12 [10]. These experiments have shown that, in particular, when resistance to uniaxial compression is approximately uniform, marble differs from limestone in having a much greater energy capacity for disintegration ($3 \cdot 10^4$ kg-m/kg as against 0.6-$0.7 \cdot 10^4$ kg-m/kg). Similar results were obtained at the All-Union Scientific-Research Institute of Drilling Techniques during investigations on explosive yielding* of rocks having various properties [16]; the results indicate that the effects of electrical discharge and of explosions on solid bodies are alike.

---

* According to M. A. Sadovskii [15], the explosive yielding of a solid medium is characterized by the relation of depth of funnel-shaped disintegration to the edge of the applied explosive charge (of cubic form).

Laboratory studies on the process of forming holes by electrical discharges in liquid were made on blocks of rock. Various models of drills were made for these investigations. Best results were obtained from a "tangential" drill, in which the axis of spark discharge was directed approximately along a tangent to the circumference of the bottom.

Discharges were produced alternately between two pairs of neighboring electrodes. The zone of most intense shattering of the rock at the bottom acquired a polygonal form. The process of shattering the rock began with the development of a network of fine fractures, which then rapidly expanded and deepened, and this permitted spalling of comparatively large rock fragments.

If the surface of the block is insufficiently cleaned of disintegrated rock particles, the effectiveness of the discharges rapidly decreases, since a considerable part of the energy is expended on comminution. In addition, the presence of mud on the shattered surface leads to dissemination of the energy of the shock wave traveling through the liquid. For the hole to acquire a form that is approximately round, three or four pairs of electrodes arranged about the periphery are sufficient. The rock occurring within the polygon, as a kind of core, is easily shattered by disturbances arising from discharges in the liquid.

Figure 13 shows the face of the operating part of a drill after a test [10]. On the ends of the individual pipes one may observe fractures; these indicate the direction of discharge from each positive electrode toward two negative electrodes.

At an energy of discharge of 100 kg-m, the expenditure of energy, according to experimental data, amounted to 143 kg-m/cm^3; i.e., one discharge disintegrated 0.7 cm^3 of rock (for rock with a crushing strength of 250-300 kg/cm^2).

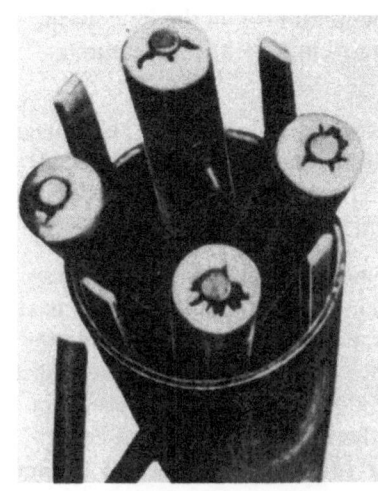

Fig. 13. Terminal (operating) part of drill from the electrode end.

Experimental investigations have shown that in the process of producing holes by explosions in models [16] and under actual conditions [17], the expenditure of energy required in experiments on samples of rocks, as blocks of various sizes, and on outcrops of rocks on the ground surface does not correspond to the actual energy capacity of disintegration for the same rocks in their natural mode of occurrence of great depths.

In the latter position, drilling (or explosive yielding) is poorer in most rocks because of compaction by hydrostatic, formational, and rock pressure [18], which increases the expenditure of energy three to fivefold.

In analogy with explosions, one should expect diminished effectiveness of discharges in plastic rocks.

The use of electrical discharges in liquid for disintegrating rocks is of interest because of the simplicity and precision of regulating the principal parameters (intensity and frequency of discharge).

The production of an on-bottom arrangement for drilling deep holes by using electrical discharges in liquid requires the development of an immersible high-voltage condenser suitable to the diameter of the hole. The solution of this difficult technical task is complicated by the transition, in drilling deep oil and gas wells, to smaller diameters (200 m and less). In regard to the transmission of powerful pulses of current of short duration from the ground surface to the bottom of a hole, because of the difficulty mentioned, the use of the electrohydraulic effect in drilling may be apparently limited to relatively shallow holes [11].

Explosions. The purpose of the task confronting us is to increase the effectiveness of sinking deep holes by developing a controllable explosive process of shattering hard rocks without the participation of any mechanical rock-crushing instrument.

Above we described a process of breaking rock by high-pressure impulses of liquid jets, ejected from the chamber of a water-jet device during the combustion of a self-igniting liquid fuel mixture. The development of this idea led again to the use of chemical energy for breaking rocks, but rather by explosion of mixtures similar in composition to the preceding. In this new process, the oxidizing agent and the fuel, present in proportions near their stoichiometric ratios, form a liquid explosive mixture, whose detonation may be effected by a chemical

17

initiator explosive. The explosion of each charge disintegrates a certain volume of rock, which is removed from the hole by flushing. By setting off a sequence of explosions a hole is formed, the diameter of which, with underwater explosions, is considerably larger than the size of the explosive charge.

During explosive drilling various methods may be employed to place the charges of liquid explosives or their components on the bottom of the hole. For example, the liquid oxidizing agent and fuel may be introduced into the hole in special capsules, giving the charge a definite form an and fulfilling a number of other purposes. Below we discuss another variant, by which the charge of liquid explosive is placed on the bottom of the hole without being encased [19, 20]. The principal rules of underwater explosions and the nature of their effects on the surrounding medium are the same in both processes. Both are also free of danger in operation, an advantage involving the fact that the charges of explosive mixture are formed directly in the hole during the process of drilling, being prepared from severally nonexplosive liquid oxidizing agents and fuels.

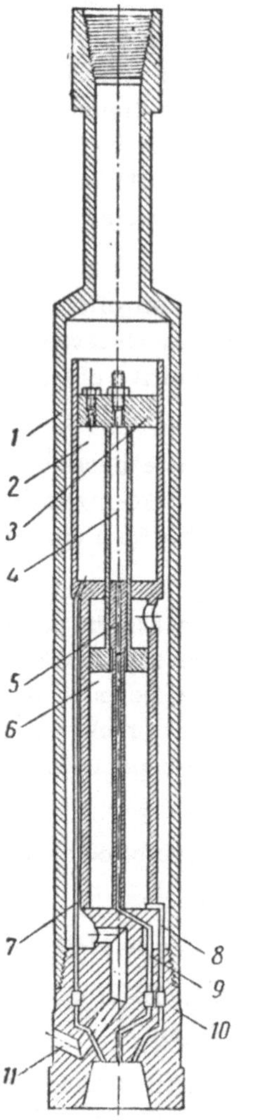

When using the capsule variant of explosive drilling, as will be shown in Chapters 7 and 8, the simplest of instruments is employed; it is placed immediately next to the site of explosions on the bottom of the hole.

During explosive drilling with uncased charges, a rather complex apparatus is placed on the bottom of the hole immediately next to the site of explosions; this apparatus is necessary for preparing the charges and for initiating them. Types of apparatus for explosive drilling differ in the method of measuring out the liquid components of the explosive mixture. Some may effect this with an independent drive for the dosage measuring device (an electric motor or a hydraulic turbine), and some effect it by automatic measuring of charges during the explosive process.

Figure 14 shows the principal features of an explosive apparatus with automatic dosage measurement of charges; this apparatus was designed by A. A. Pavlichenko. In it, all the components (oxidizing agent, fuel, and chemical initiator explosive) are supplied continuously, but the charges are measured out periodically by the effect of the pressure arising during the explosions.

The apparatus operates in the following way.

During movement of the two-stage piston (because of drop in pressure during the pumping of flushing liquid through the apparatus), the components of the explosive mixture are fed through the pipes into the mixing sprayer, and they flow out of this opening into the space beneath the lead. The form, cross section, and disposition of the channels in the sprayer are so chosen that the jet of chemical initiator encounters the jet of explosive mixture when the liquid charge reaches the desired volume and impinges on the solid surface to be disintegrated. At this instant a detonation occurs. A jump in pressure in the space beneath the head, occasioned by the explosion, interrupts the supply of components, but the supply is renewed after a drop in pressure. As the procedure continues the explosions and "cut-offs" are repeated automatically. By modifying the construction of the head and, in particular, its mixing arrangement (a description of these adaptations is omitted), it has been possible to regulate the size of individual charges (from 3 to 20 g) and the frequency of explosions (from 150 to 2500 per min) within suitable ranges.

Fig. 14. Diagram of an explosive apparatus with automatic dosage measurement of components of liquid explosive mixture and of the chemical initiator. 1) Case; 2) oxidizing agent; 3) two-stage piston with cavity for chemical initiator; 4) chemical initiator; 5) plunger with valve for passing the chemical initiator; 6) fuel; 7) pipeline for oxidizing agent; 8) pipeline for fuel; 9) pipeline for chemical initiator; 10) head; 11) conduit for flushing liquid.

The compositions of the liquid explosive mixtures of uncased charges are generally similar to those of the charges in capsules, except that there is a more limited choice of oxidizing agents and fuels because of their direct contact with the flushing liquid in the hole. In addition, a eutectic alloy of alkali metals, which reacts violently with water and the oxidizing agent, is used as the initiator for uncased charges, whereas other methods (besides chemical initiation) may be used for setting off a charge in a capsule.

Many small holes with diameters of 200-350 mm have been drilled in limestones by explosive devices (from the ground surface). The explosive drilling process has been carried on chiefly in air with the charges placed in the central part of the bottom of the hole. From holes that were filled with water or clay muds the liquid was thrown out by gas-forming products of the explosions; mud and large rock fragments were ejected along with the liquid.

Experiments in a hole about 200 m deep in which flushing liquid circulates have not yet demonstrated the possibility of using the explosive drilling process with uncased charges of explosives. Furthermore, the experimental data thus far obtained (for explosions in air) point to comparatively low effectiveness of explosions of simple uncased charges, and this effectiveness is diminished yet more when a certain frequency of explosions is attained.

Experimental data have shown that when an explosion occurs along the length of a set charge of liquid explosive, initiated by an igniter situated above the charge, the penetration into the rock is substantially greater than the penetration from an explosion of a flattened charge of uncased liquid explosive of arbitrary form, set on the bottom of the hole and having an undetermined position for initiating the explosion.

The decrease in effectiveness of explosives with increase in frequency is explained by the intense elimination of incandescent gases, in which the oxidizing agent and the fuel are partially burned or detonated; this prevents the formation of a completely proper mixture and the possibility of achieving contact with the solid face to be shattered.

Cleaning the bottom of the hole of shattered rock has proved to have a great influence on the effectiveness of the explosions. When the bottom is improperly cleaned, fragments of various sizes accumulate and shield the bottom from the effect of the explosion.

Better cleaning of the bottom is achieved during washing by using a high-speed pressurized stream of washing fluid. But this technique greatly complicates the production, the placing on the bottom, and the protection from erosion of exposed (uncased) charges of liquid explosive. The protection of the uncased explosives from erosion may be achieved by circulating the washing fluid at some distance from the bottom. By doing this the disintegrated rock is partly removed from the bottom in the circulating stream of the flushing medium by the explosive disturbances transmitted through the liquid or by gas-forming products of the explosions (when they are frequent). However, as experience has shown, failures of explosion (especially when the apparatus is being lowered) in combination with the absence of bottom flushing lead to accumulation of explosive on the floor of the hole. The discharge of the accumulated explosive when the chemical initiator first strikes it may destroy the apparatus.

The development of an explosive drilling process on the basis of charges of uncased liquid explosive, despite the difficulty of effecting it, is expedient, chiefly for sinking holes at deeper intervals in hard rocks. The process may be effected by using inserted devices, without raising the pipe from the hole. An important advantage of this method is the high frequency of explosions.

* * *

Of the impulse processes of shattering rocks, the highest development has been attained for underwater explosions of liquid explosives within capsules.

The results of development and investigation of this process form the basic contents of the present book, of which Chapter 2, in which the problems of forming holes by explosion are discussed, may be of general interest in regard to the mechanism of shattering rock subjected to pulsing charges.

B. Thermal Processes of Shattering Rocks

Above we have discussed some wave processes which are capable, under certain conditions, of shattering rocks by impulses of high pressure under the combined effect of compressional and rarefaction waves propagated from the free surface (bottom of the hole) into the depth of a mass of rock.

For rocks the tensile strength is much less than the compressional strength (from one-fifth to one-fiftieth as great [21]). Therefore, the zone of shattering acted on by a compressional wave may be increased during the passage of a rarefaction wave. In this connection it is advisable to search for such methods that the effect on the rocks during their shattering be accompanied by spalling of large fragments through the action of tensional and buckling forces.

The relative value of the parameters of compressional and rarefaction waves affecting the shattering of rocks by impulses of high pressure has been inadequately studied; however, if factors are at work to repress the rarefaction wave, the intensity of shattering should apparently decrease. One of such factors may be shown to be hydrostatic pressure in a hole filled with liquid.

It is well known that the result of impulses of high pressure acting on rock depends on the physical-mechanical properties of the rock, and these properties change markedly with depth of occurrence of the rock. Therefore, in regard to most deep-lying rocks, which have been compacted by high rock and hydrostatic pressure and in which drilling by any known mechanical means becomes markedly more difficult with depth, it is expedient to use, under these conditions, a more effective technique, thermal action, to the extent of fusing and even vaporizing the rock-forming minerals.

To solve many scientific and technical problems, ever higher temperatures are required; the attainment of these temperatures may be possible by using chemical, mechanical, electrical, and nuclear energy. At present there are no stationary long-acting sources of high temperature employing nuclear energy. The upper limit of temperature obtained by chemical reaction is about 5000-6000°K. Higher temperatures may be obtained in the channel of an electric arc (at atmospheric pressure).* Of interest is the method of increasing the temperature of an arc by forceful restriction of its diameter and by obtaining a high-temperature jet of plasma.

Thermal shattering of rocks may occur in various ways:

Shattering as a result of thermal strains set up in the surface layer of the rock; shattering of this type may be produced by intense heating of the rock (such as by a flame or by gases of a jet burner), the heat gradually disseminating from the site of heating by thermal conductivity;

shattering of rock by thermal strains arising during simultaneous heating of several zones from a free surface within the mass; such heating may be effected by high-frequency currents;

lastly, shattering of rocks, possibly, by fusion or vaporization of the rock-forming minerals, such as by high-temperature gas streams from a jet burner, by high-frequency currents, or by high-temperature streams of plasma.

The intensity of the process of fusion or vaporization of rocks apparently depends little on the mechanical properties of the rocks, particularly on the plastic state, which is found in rocks that occur at great depths. Therefore, from a still preliminary point of view, it is possible to foresee that thermal processes of shattering some deep-lying rocks may find preference over mechanical processes, including impulses of high pressure.

Gas Jets. The possibility of using high-temperature gas jets for shattering rocks is attested by extensive literature, such as [22, 23, 24, and 25]. The mechanism of shattering rocks by heat depends, in this case, on the unequal expansion of individual crystals and on considerable stresses in the medium because of temperature gradients.

The principal factors determining the effectiveness of shattering are the hardness and elasticity of the rock, thermal conductivity, thermal capacity, and coefficient of thermal expansion of the individual minerals and of the rock.

The lower the modulus and the higher the limit of elasticity in rocks, the more difficult it is to produce shattering by heat. Rocks having a low thermal conductivity and varying thermal capacities for the constituent

---

* In superpowered freely burning arcs, it is possible to obtain temperatures no greater than 10,000°K.

20

minerals and having a high temperature gradient are shattered more easily. The most intense shattering is observed in rocks having a coarse-grained texture and a distinctly complex mineral composition.

In hard rocks, even when the change in volume is slight, large tensional stresses are developed, producing more effective shattering than in soft rocks.

Thermal drills, producing gas jets, consist of a burner and an arrangement for underwater fuel (kerosene), an oxidizing agent (gaseous oxygen), and water. Successive treatment of the bottom by high-temperature gas jets is achieved by suitable distribution of nozzles and by rotation of the thermal drill.

The temperature in the combustion chamber of a thermal drill depends on the ratio of oxidizing agent to fuel, and this amounts to about 2500°K at a pressure in the chamber of 5-7 atm. Particles of rock that are broken away from the mass are removed from the hole by currents of gas, the velocity of which on emerging from the nozzle of the burner reaches approximately 2000 m/sec. The combustion chamber and the burner of the thermal drill are cooled by water, which, on meeting the streams of gas, is converted to steam, facilitating the removal of disintegrated rock particles from the hole.

According to experimental data, the expenditure of oxygen in drilling shallow dry holes 150-250 mm in diameter at a rate of 4-6 m/hr reaches 250-300 mm^3/hr (250-300 · 10^{-9} m^3/hr); this constitutes 60-70% of the direct outlay in drilling. To decrease this expense, it is possible to use a thermal drill operating on liquid fuel and air enriched in oxygen.

The use of high-temperature gas jets for shattering rocks, the jets being obtained by burning a mixture of liquid fuel and gaseous oxygen, is limited to use in the dry sections of a hole at depths of some tens of meters, chiefly in crystalline rocks.

The use of such a process for drilling deep holes is possible in principle, but the flow of heat must be sufficient to fuse the rocks in a time interval substantially small in comparison with the rate of heat transfer in the

Fig. 15. Products of shattering granite, fused by jets of gas from the combustion of liquid oxidizing agent and fuel in a jet burner.

rock. This is explained by the various properties of the rocks alternating in the section of a deep hole. The shattering of many of these rocks merely through thermal stresses may prove to be ineffective or impossible.

To sink a hole with a diameter of 150 mm at the rate of fusing siliceous rocks at 10 m/hr, the expenditure of heat, according to approximate computed data, would amount to 3000 kcal/sec, which corresponds to a specific heat flow at the bottom of the hole of 1700 kcal/m^2 sec. Such a flow might be obtained by burning high-caloric fuel (based on liquid oxidizing agent and fuel) in the chamber of the jet burner (reproducing the design of a liquid jet engine).

Experiments on breaking down blocks of various rocks by high-temperature gas jets, streaming from the nozzle of a jet burner with supersonic speed, have been conducted with an expenditure of fuel of about 1 kg/sec. In these experiments, especially with granites and ferruginous quartzites, a high rate of penetration was achieved (30-40 m/hr) with the formation of a cylindrical opening 250-400 mm in diameter. The products of destruction of the granite (Fig. 15) form congealed, highly porous and brittle masses, which are broken down and blown out of the opening in the block by the current of gas.

To carry out this process in a deep hole it is necessary to solve a number of complex technical and technological problems occasioned by the high hydrostatic pressure, the influx of formational water, the removal of breakdown products from the hole, and other factors.

<u>Plasma</u>. Gases heated to high temperatures, by an electrical discharge for instance, become ionized. It is known that, for a certain degree of ionization, the gas acquires a peculiar status, being converted into plasma, which consists of free electrons and positive ions formed by the tearing away of one or several electrons from the outer electron shells of the atoms.

It is possible to use plasma for vaporizing and fusing rocks as a means of drilling holes.* To accomplish this it is proposed that an electrical discharge be produced in a manner that will form high-temperature plasma, in a special gas-discharge chamber (plasma generator) on the bottom of the hole. Jets of plasma will stream with supersonic speed (5000-6000 m/sec) against the bottom of the hole, causing melting and vaporization of the rock.

During the reaction between the high-temperature jets of plasma and the rock one may observe a zone in which fusion and vaporization of the rock-forming minerals occur and a zone of thermal stresses in which the rock cracks and in which particles of various sizes are separated from the mass.

The products of shattering are removed from the hole by the ordinary method, i.e., by circulating streams of flushing liquid, gas, or steam. The gaseous products from the zone of vaporization are condensed in the flush-

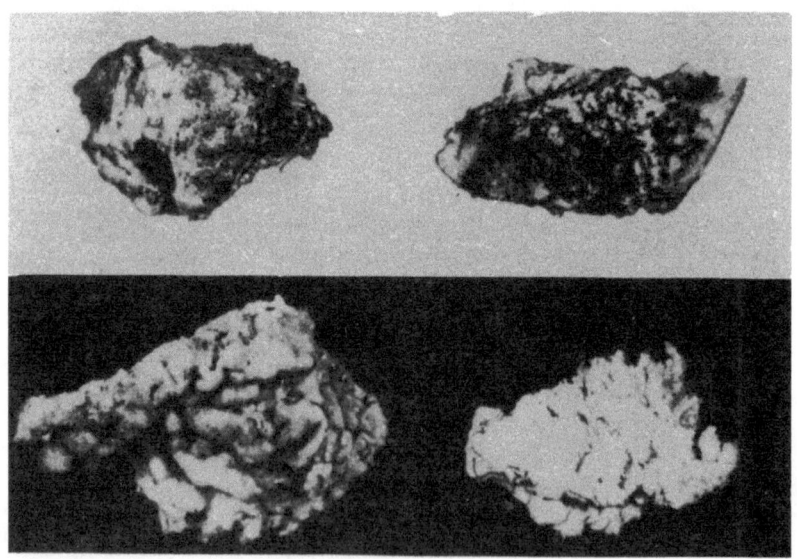

Fig. 16. Breakdown products of cherty limestone (above) and granite (below), fused by high-temperature jets of plasma.

ing liquid and precipitated as small solid particles. The products derived from the zone of fusion, at a temperature of 15,000-20,000°C in the path of the discharge arc, form a brittle mass, the broken particles of which are of various sizes and shapes, as shown by experiments on cherty limestone and granite (Fig. 16).

---

*Proposed by the author conjointly with N. V. Aleksandrov, V. K. Bogomolov, A. I. Gol'binder, N. I. Strizhov, and A. M. Khazen.

It is clear that when the bottom of the hole is washed with a liquid having a high thermal capacity, only a small part of the heat may be transferred to the rock.

In order to decrease heat loss it is necessary to divert the stream of flushing liquid so that less of it is directed against the bottom and is used to cool the discharge chamber. The amount of liquid used for cooling is such that when vaporization of the liquid occurs at the bottom, bubbles of steam will be preserved. The main bulk of the liquid circulates during this time above the discharge chamber (upper flushing).

It is assumed that the breakdown products are removed from the bottom by streams of gas and steam, streaming with supersonic speed, introduced into the circulating flow of flushing liquid, and carried to the surface.

In experiments on fusing and vaporizing granite and cherty limestone, jets of high-temperature plasma produced a cylindrical hole, the walls of which were formed of a glass-like crust several millimeters thick.

The Established Design for an Arc Plasma Generator.* The basic parameters of a gas in the plasma state are the degree of ionization, the temperature, and the pressure. The degree of ionization, and also the mobility of the electrons and ions, depends on the resistance of the gas-discharge gap.

In determining the established design for an arc generator of plasma, it is possible to set up equations, showing the relationship between the degree of ionization in the zone of discharge and the temperature and pressure and showing the relationship between the temperature and pressure of the gas and the balance of energy in the discharge and the balance of mass in the discharge. For the relation of degree of ionization to temperature and pressure, the equation of Sakh [26] is used, defining the concentration of charged particles $x = n_i/n_0$, by which is established the dynamic equilibrium in the quantity of ionized particles (n_i) and neutral atoms (n_0):

$$\frac{x^2}{1-x^2}\, p = 2.4 \cdot 10^{-4}\, T^{2.5}\, e^{-\frac{qv_i}{kT}},$$

where p is the pressure at the given temperature (in millimeters of mercury), T the absolute temperature in degrees Kelvin, qv_i the work of ionization in ergs, and k the Boltzmann constant in ergs per degree.

Figure 17 shows a graph based on the equation of Sakh, representing the relation of degree of ionization to temperature at pressures of 1 and 400 atm. The graph shows that with an increase in pressure, for a given

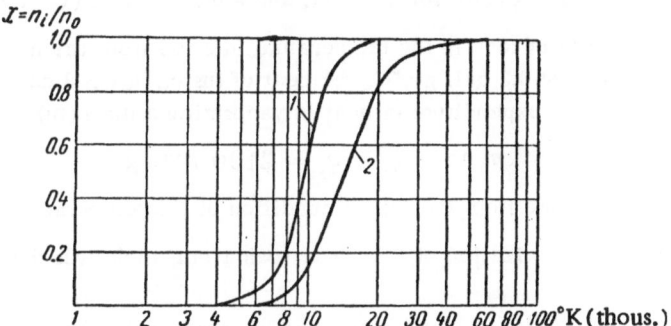

Fig. 17. Relation of degree of ionization to temperature and pressure in the channel of arc discharge. 1) For pressure of 1 atm and work of ionization of 7.5 ev; 2) for pressure of 400 atm and work of ionization of 6.0 ev (1 ev = $1.602 \cdot 10^{-12}$ erg).

temperature, the degree of ionization diminishes, and this permits a higher temperature, for the same expenditure of energy, to be obtained.

The relation of the pressure p and temperature T to the computed ionization may be obtained (for argon) from the modified Clapeyron equation [27]:

$$p = \varrho R (T_g + T_e),$$

where ρ is the density of the gas, R the gas constant, T_g the temperature of the gas, and T_e the electronic

---

*Written in cooperation with A. M. Khazen.

temperature (in the column). For a polyatomic gas one may use the tables of thermodynamic functions for air of A. S. Predvoditelev and others.

The equation characterizing the balance of mass and the balance of energy is founded on experimental data and the design parameters of the plasma generator.

The equation for the balance of mass may be written:

$$M_1 + M_e = M_g,$$

where M_1 and M_e represent the masses, respectively, initially in the arc path, derived from vaporization of the cooling liquid, and from destruction of the electrodes; M_g represents the mass of gas going into the formation of the jet of plasma.

The balance of energy in the discharge with a supplied power of UI is determined by the equation:

$$UI = S\,\alpha\sigma\,T^4 + Q_e + \frac{M_g}{\mu}\left(\int_{T_0}^{T} C_p dt + Nxqv_i\right),$$

where S is the area of the surface of the arc column, α the degree of blackness of the discharge, σ Stefan's coefficient (equal to $5.75 \cdot 10^{-12}$), Q_e the amount of heat lost at the electrodes, μ the average molecular weight of the gases, N Avogadro's number, and C_p the thermal capacity of the gases at constant pressure.

In the equation for balance of energy, the first member on the right corresponds to losses through radiation, the second to losses at the electrodes, and the third to energy imparted to the gaseous medium.

The cited equation permits an approximate determination of the temperature, pressure, degree of ionization, and density of gas within the chamber of the arc generator of plasma at any given power supplied to it and any given amount of heat necessary to break down the rock.

The computation of the velocity of emission of a jet should be made according to the parameters obtained for the chamber and for a certain external pressure. For an approximate computation of the quantity of heat necessary to break down the rock (silica, for example) by vaporizing it, we shall assume a diameter of 15 cm for the hole, a density (ρ') of 3 g/cm^3 for the rock, and a drilling rate (v) of 10 m/hr.

The average thermal capacity of rock in the temperature interval from the initial temperature to the temperature of fusion (2700°C) is $c \approx 0.3$ cal/g · °C; the heat of fusion (q_1) is 3 cal/g, and the heat of vaporization (q_2) is 1300 cal/g. The total expenditure of heat in vaporizing a mass (m) of 1 g of silica is

$$Q_1 = cm\,\Delta t + q_1 + q_2 = 2100 \text{ kcal/g},$$

where Δt is the increase in temperature from the initial condition of the temperature of vaporization.

The expenditure of heat on vaporization of a layer or rock 1 cm thick over the entire cross section of the hole is

$$Q = Q_1 \rho' \frac{\pi d^2}{4} = 1.12 \cdot 10^6 \text{ kcal} = 4.68 \cdot 10^6 \text{ joules}.$$

Hence the power necessary to be transmitted to the rock from the gas-discharge chamber, after computing losses, is

$$N = \frac{Qv}{3600 \cdot 1000} \approx 1300 \text{ kw}.$$

Let us determine the approximate value of power supplied to the plasma generator.

The losses in the arc generator of plasma are determined experimentally and depend on its construction. Under favorable circumstances the losses do not exceed 20% [27], including losses through radiation (according to the same source of data), which are about 5% at a pressure near atmospheric pressure in the zone of discharge.

With increase in pressure the losses through radiation may rise considerably; therefore, in designing arc generators of plasma for operation in a hole, it is advisable to make provisions for the rock to absorb most of the radiant energy.

With the indicated losses in the arc generator for plasma and the use of 60% of the plasma jet [28] for breaking down the rock, up to 50% of the energy supplied to the plasma generator will be expended. Consequently, the power that must be supplied to the gas-discharge chamber, with due consideration to the losses, is 2600 kw for vaporizing the rock over the entire cross section of the hole at a rate of 10 m/hr.

Corrosion of Electrodes. It is of special importance to consider the problem of stability of electrodes. The temperature of the anode and cathode in most cases remains at the level of the boiling point, the values of which for several materials are shown in Table 1 [29].

The temperature of the anode in the zone of high-pressure* discharge is greater than the temperature of the cathode, and its corrosion is therefore more intense.

TABLE 1. Boiling Point of Some Materials in Electrodes

Composition of electrode	W	C	Pt	Fe	Cu	Al	Hg
Boiling point, °K	4830	4000	3800	2500	2300	1800	350

The boiling point of the electrodes in relation to pressure in the zone of discharge is shown in Table 2 [29].

From Table 2 it follows that the stability of the electrodes should increase with increase of pressure in the gas-discharge chamber. However, for prolonged arc discharges the process of electrode destruction occurs by oxidation of the material and is independent of the temperature. There is an optimum concentration of oxygen in the gaseous medium that corresponds to minimum corrosion of the electrodes. A low concentration of oxygen is not always desirable, since, under such circumstances, some materials are covered with a protective film of oxides.

An important feature of arc discharge is the basic possibility of its occurring with cold anode and cathode only through the emission of electrons from the near-cathode zone of burning gas and through thermal ionization. This may be of practical value in increasing the stability of the electrodes.

TABLE 2. Relation of Boiling Point of Electrode Material to Pressure

Pressure in zone of discharge, atm	0.1	0.5	1.0	2.0	10.0	20.0
Boiling point of electrode material, °K*............	3940	4145	4200	4900	6520	7560

*Electrode material = tungsten and admixtures.

According to experimental data, a copper electrode (anode), cooled by water at a temperature of 16,000°K in the arc channel (current of 225 amp), deteriorates at the rate of 0.8 mm/min.

Design of a Plasma Generator. The most effective means of increasing the temperature in the column of the electric arc is the limitation of the diameter of the column by active removal of heat from the periphery. In this way the arc cannot increase in diameter when the current strength increases. As a consequence, the current density, the power emitted per unit volume of arc, and the temperature in the column increase.

Figure 18 shows a diagram of a plasma generator. By using such an apparatus one may obtain a jet of plasma in the temperature range of 8000 to 17,000°K (at atmospheric pressure) and may maintain it long enough to perform his experiment [30].

---

*For electrical discharge in gas at a pressure of 1 atm and more.

A limitation of the diameter of the discharge column may also be obtained by stabilizing the diaphragm by a cooling film of water.

A photograph of a high-temperature high-velocity gas jet is shown in Fig. 19. The average diameter of the arc is 20 mm, the length 175 mm. The arc burns between cooled carbon electrodes. The stabilizing diaphragm has an opening 3 mm in diameter.

An arc stabilized by water has been described in the literature [31]. The charge in this arc is very great. The diameter of the arc column is 1.4 mm, the length 13 mm, the current 250 amp, the potential of the field in the arc column 240 v/cm, the power per centimeter of length in the arc 60 kw, and the power per cubic centimeter 1300 kw. The temperature on the axis of the arc is 35,000°K. During observations on the axial symmetry of the entire arrangement (electrodes, sprayer-diaphragms) the arc burned steadily.

Data from another paper [32] indicate that the maximum temperature at the axis of the arc has been proved to be 55,000°K, the power, referred to the length of the arc, 435 kw/cm, and to the volume of the arc, 10,500 kw per cm^3.

The conversion of electrical energy to thermal energy in the stream of plasma permits the attainment of high specific power at the bottom of a hole of minimal diameter.

The attainment of a high-temperature gas jet by electrical methods has both advantages and disadvantages in comparison with chemical processes of similar design.

Advantages of the electrical methods are the possibility of obtaining a greater temperature than by burning fuel and the achievement of a high degree of control.

At the same time, the procurement of high-temperature jets of gas and plasma by electrical methods is accompanied by relatively weak formation of gas, insufficient for removing the breakdown products from the bottom of the hole. It is therefore necessary to introduce a considerable quantity of supplementary liquid or gas.

An investigation of the problem of using high-temperature jets of plasma for drilling deep holes under high hydrostatic pressures should precede the study of peculiarities of the production of arc discharges under these conditions.

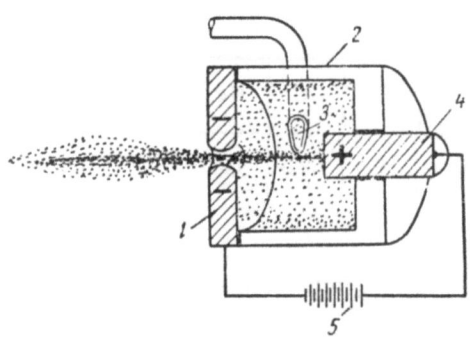

Fig. 18. Diagram of a plasma generator with orificed cathode. 1) Orificed cathode; 2) housing; 3) conduit for cooling agent; 4) anode; 5) source of direct current.

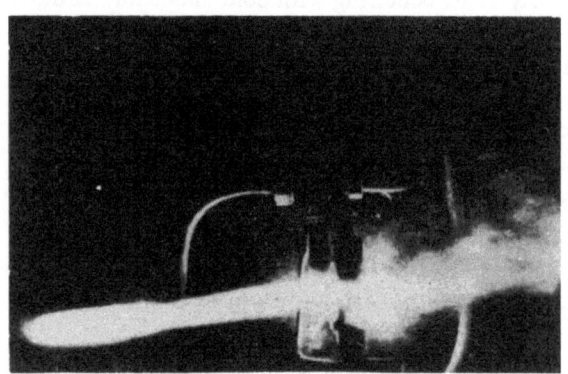

Fig. 19. Tongue of plasma beyond the cathode.

High-Frequency Currents. A high-frequency current will heat a rock simultaneously throughout its entire volume, piercing the rock by a field of force lines. The rate of heating is substantially faster than local heating by an open flame or streams of gas from a jet burner, and when the temperature reaches 150-300°C the temperature gradient is found to be sufficient to form, in some rocks, macroscopic fractures of various depths and lengths; microscopic fractures will also be formed, increasing the brittleness of the rock. To accomplish such shattering by ordinary thermal methods a considerably higher temperature is required.

The conversion of the energy of the electromagnetic field to heat may be accompanied by losses, because of various factors:

Reversal of magnetism (hysteresis), if the rock possesses ferromagnetic properties;

Eddy currents (Foucault currents) in rocks with well-defined conductivity;

Mutual friction of neighboring polar molecules during rotation as they respond to the direction of the electrical field in the rock dielectrics;

Ionization of gases included in the rock. ,

Consequently, depending on the properties of the rock, it is advisable to use waves of different ranges and to use various methods of transmitting the electromagnetic energy.

For rocks with marked magnetic properties it is suitable to use an induction method, by which heating is produced through alternation of the magnetic field in a coil supplied with high-frequency current. Dielectrics may be heated in an electric field between the faces of a condenser or of a special antenna. The rock is shattered because of unequal heating in minerals with different electrical characteristics and with different coefficients of linear expansion.

The generation of thermal stresses is also possible when a rock with uniform electrical properties is heated by concentration of the electromagnetic field in some small zone.

This type of shattering represents a variety of mechanical activity on the rock, but is distinguished from ordinary mechanical processes by the creation of stresses not by applying force to the surface of the material, but rather to some zone extending into the interior of the mass.

During high-frequency heating it is also possible to fuse and vaporize the rock-forming minerals, but to accomplish this the power supplied to the rock would have to be substantially increased. In addition, in rock dielectrics, the heating of which occurs in the electrical field of an antenna, incipient fusion of the rocks ceases because of a sharp rise in their conductivity and through reflection of a considerable part of the wave energy.

In magnetic ores heated in high-frequency magnetic fields this phenomenon begins on reaching the Curie point, i.e., the temperature at which the material loses its magnetic properties.

Experiments have been performed on cracking blocks of iron ore in a magnetic field of 100-200 oe and a frequency of 240 kc [33]. In some experiments the samples were placed entirely in a homogeneous magnetic field; in others the magnetic field, concentrated within a coil, heated only a small part of the sample (the coil had an inner diameter of 9 cm). In the latter experiments, because of the large temperature gradients the thermoelastic stresses were increased; $\sigma_T = \alpha\, ET$ (α is the coefficient of thermal expansion and E is Young's modulus). Thus, for ferruginous quartzites from the Kursk magnetic anomaly ($\alpha = 1.1 \cdot 10^{-5}$ and $E = 1.2 \cdot 10^6$ kg per cm^2), at $T = 300°C$, $\sigma_T = 4000$ kg/cm^2, which is greater than the temporary crushing strength ($\sigma_{cr} = 3300$ kg/cm^2) and exceeds by many times the tensile strength and shear resistance.

When frequencies are increased to the order of 10^8-10^{10} cps for intense heating of a limited zone, it is possible to transmit high-frequency energy to the rock by a directional antenna such as the horn or other radiator type.

Intense shattering of blocks of nonmagnetic rock during heating by an electrical field is greater the higher its dielectric permeability and its loss-angle tangent. The greater the frequency the smaller the zone of heating and more intense the evolution of heat.

The power evolved in the rock may be determined by the formula [34]

$$P = 5.55\ \varepsilon\ \text{tg}\ \delta\ f E_1{}^2 \cdot 10^{-7}\ \text{w/cm}^2,$$

where ε is the relative dielectric permeability, δ the loss angle, f the current frequency in cycles per second, and E_1 the potential of the field in the rock in kilovolts per centimeter.

As with a magnetic field, the absorption of energy of the electrical field by ever possible micro- and macroinclusions in the rock is variable; because of these, thermal stresses are concentrated and the brittleness of the rock is increased.

The exposure of sandstone, basalt, hornfels, and other rocks to high-frequency (50 Mc) electrical fields and radio waves in the centimeter range (3000 Mc) generated by a magnetron substantially increases the brittleness of the rocks, and thus facilitates the breaking of the rocks by some mechanical device [35].

For drilling purposes there is interest in the initial heating of the rocks and the concentration of thermal stresses at some distance from the free surface. In such a process there occurs something like a thermal impact within the exposed mass, fostering the breaking off of rock fragments, which may then be broken by a mechanical device.

Two procedures for shattering rock in this manner may be noted.

In one of these methods a standing electromagnetic wave is produced in the rock; this may be accomplished by using a reflecting shield (such as a sheet of copper). This technique has been used to break up large blocks of ore (oversized) and nonmagnetic rocks, or, as an example, masses of ice. The reflecting shield in this latter example was water, lying beneath the layer of ice [33].

According to the other method a combination of several radiators is used, the energy of a high-frequency electromagnetic field being focused at some depth beneath the layer of rock, which is heated considerably less than the zone where the energy is focused.

From the patent literature a design for drilling holes has been found by which the rock is shattered by the energy of ultrahigh-frequency currents and the bottom is cleaned by a current of air or gas. In this design a magnetic generator is placed on the surface and the energy is transmitted to the bottom of the hole through a waveguide, which, it is suggested, may ordinarily be a column of drill pipe. At the end of the waveguide channel (pipe) a radiating system is placed, in the form of a horn or a multislotted type.

The functioning of the indicated method when sending high-frequency energy great distances, as encountered in deep oil and gas wells, becomes difficult because of great loss. According to existing data [36], when a waveguide of copper pipe 56 mm in diameter is used, prepared with sufficient precision (as to variations in wall thickness and ellipticity), the loss, measured by the ratio of power supplied the waveguide to power evolved at the attached charge, amounts to as much as 1300% per kilometer of waveguide at a frequency of 24,000 Mc. Therefore, even with markedly lower frequencies and losses, the use of high-frequency currents for drilling deep holes is more suitably effected by placing a magnetic generator near the bottom of the hole. However, under such circumstances, if the hole is filled with water, the efficiency of the system remains low. This fact is explained by the extensive reflection of the electromagnetic waves from the surface of the water, reaching a value of 0.7 [37], and by the absorption of the remaining part of the energy by the water. The coefficient of reflection for dry soil is 0.4 (according to data from the same source).

The efficiency of modern magnetrons in the centimeter range of wavelengths is 0.6-0.3 [38], and the losses in the current transformer and the cable of the feed system (3000 m) is computed to be 20%.

The approximate efficiency of the generator of high frequency together with the current transformer and the 3000 m of conductor is 0.5 for a magnetron of maximum efficiency (0.6) and 0.25 for a magnetron of minimum efficiency (0.3). With due consideration to the loss caused by reflection of the electromagnetic waves from the rock (0.4), we obtain values of 0.3 and 0.15, respectively, for the total efficiency when operations are in a dry hole.

In a hole filled with circulating clay mud, heating of the rock may prove to be impossible because of losses associated with absorption of the electromagnetic waves.

Thus, the use of high-frequency currents will probably be limited at first to the sinking of comparatively shallow water-free holes in rocks that are subjected to splitting. The breaking off of rock particles from the mass will be accomplished by thermal stresses resulting from concentrations of electromagnetic energy in definite (small) zones of the material to be broken. The process of exposing the rock to high-frequency currents will be combined, for intelligent operation, with the mechanical work of a drill, the operation of which will be greatly improved by the preliminary shattering of the layer of material.

*   *   *

From this survey of several processes of shattering rocks by impulses of high pressure of various intensities and durations and of several thermal processes for shattering rocks, the following conclusions may be made.

1. The shattering of brittle hard rocks by impulses of high pressure is accompanied by spalling of comparatively large fragments at relatively low specific expenditure of energy. Of the number of investigated impulse processes, the shattering of rocks by high-pressure jets of liquid, by impulses of pressure in liquids by collapse of vacuum bubbles, by electrical discharge in liquids, and by explosions satisfies in principle, the requirements of sinking deep holes under high hydrostatic pressures. The shattering of rocks by acoustical waves under the indicated conditions is impossible because of the cessation of ultrasonic cavitation with increased pressures and because of the inadequate power of present-day radiators of ultrasonic waves for direct shattering.

2. The shattering of some rocks by forces generated by thermal stresses set up by heating definite zones, and thus producing irregular expansion of individual crystals because of the thermal gradient, is but a variety of mechanical crushing. It may be effected through relatively small thermal currents (from 200 kcal/m^2 · sec up) by using chemical reactions and high-frequency currents (in combination with a mechanical device) for drilling shallow liquid-free holes.

3. The fusion and vaporization of rocks may be achieved by high-temperature jets of gas obtained chemically (such as from a jet burner) or by utilization of electrical energy in a special arc discharge, with the formation of high-temperature jets of plasma. An advantage of thermal processes is the independence of the intensity of fusion or vaporization of the rock on the physical-mechanical properties of the rocks, which change markedly with depth. In addition, it is possible to strengthen the walls of the hole, during drilling, by fusing the rocks.

Chapter 2

MODEL INVESTIGATIONS ON FORMING HOLES BY EXPLOSIONS

The present chapter considers the influence of rock properties, hydrostatic and rock pressure, and some other factors on the effectiveness of explosions; it examines the stability characteristics of solid media such as rocks under explosive charges; it defines the basic parameters of underwater explosions and of the zones of influence of these explosions on rocks in holes; and a qualitative scheme of making holes by explosions is proposed.

Method of Investigation

In view of the complexity of the phenomena of an underwater explosion and its effect on rocks in a deep hole, and in view of the lack of study of the physical processes arising during such an explosion, simplifications have been assumed in the investigation of some of the problems of forming holes by explosions and, in a number of places, quantitative evaluations have been used.

It is assumed that rock is an isotropic medium in which elastic waves, arising during the explosion, are propagated. The compressive strength of the rock greatly exceeds the tensile strength; consequently, in addition to shattering by the propagation of compressional waves, the rock may undergo supplementary shattering by the passage of tensional waves.

The examination of several problems of forming holes by explosions has been combined with experimental investigations on models.

Model studies are rather widely employed in explosive practice. In selecting a method of model experimentation, the following considerations were borne in mind:

a) Because explosive drilling must be done with comparatively small charges (up to 50 g) in order to produce holes of the required diameter (200-300 mm), laboratory experiments may be carried out with comparatively small-scale models.

b) Some experiments having to do with the study of various factors (size and shape of charge, hydrostatic pressure, etc.) must be conducted on homogeneous material with reproducible properties, whereas other experiments require materials having various mechanical properties.

c) A hole is formed by the sequential effect of a great number of explosions, and the technique of conducting the experiments should be consequently rather simple and efficient.

In accordance with the indicated considerations, experimental investigations were made on rods and blocks of Plexiglas and, chiefly, on blocks of concrete molded in steel rings 210-400 mm in diameter and 400-1200 mm high. The charge employed ranged from 0.02 to 0.10 the dimensions of an actual charge. The corresponding linear scale on the models ranged from 1 : 3.7 to 1 : 2.15.

In most of the experiments an electric detonator (KD-8) of fulminating mercuric tetryl was used as a charge. Some experiments were made with small charges of liquid explosives.

In experiments on studying the effects of explosives on different materials and rocks, the samples were placed inside a steel ring which was filled with concrete. The bottoms of the model holes were cleaned by streams of water, the volume and velocity of which approximately conformed to the actual conditions encountered in cleaning the bottom when drilling an actual hole.

Rock Properties and the Effectiveness of Explosions

In order to shed light on the connection between indices of explosive drilling and rock properties [16, 17], materials of various strengths were analyzed, by ordinary static methods, when subjected to crushing by dynamic loading. The results of some of the characteristic experiments are shown in Table 3.

TABLE 3. Results of Explosive Drilling of Model Holes in Various Materials*

Test material				Penetration of explosion		Diameter of hole, mm	Average volume of rock shattered by a single explosion	
Name	Density, g/cm³	Porosity, %	Ultimate crushing strength, kg/cm²	mm	%		ml	%
Concrete	2.35	4.6	400	3.0	100	50	5.9	100
Concrete	2.32	4.8	270	3.8	127	52	8.0	135
Granite from the Sokolovskii quarry	2.69	0.6	1700-2000	3.1	103	48	5.6	95
Sandstone	2.66	8.8	400	4.2	140	45	6.7	114
Turusskii limestone	2.60	14.6	370	2.8	93	42	3.9	66
Marble from the Sadlykhlo quarry	2.71	0.8	1200-1600	1.5	50	80	7.3	124
Limestone from the Kamushka quarry	2.59	18.0	130	5.2	173	57	11.6	197

*The explosions occurred under a layer of water 6 m thick.

In examining the data in Table 3 it may be noted that for materials having values of ultimate resistance to uniaxial static compression that vary considerably (thirteen to fifteen-fold) the penetration of explosions varies only by a factor, approximately, of 1.5. However, for rocks with similar values of ultimate compressive strength (such as granite, marble, Turusskii limestone, and sandstone) the variation in penetration per explosion is considerably greater. Consequently, in relation to explosive yielding, rocks are not always disposed in the same sequence as is possible when employing mechanical means of shattering them and ordinary methods of static testing, techniques not applicable in evaluating strength characteristics of rocks when using explosive charges.

Experiments showed some effect of the properties of the material on the shape of the model hole formed by explosions.

In concrete and test rocks the segment of the hole between the bottom and the zone where a more or less consistent cylindrical shape is established has the form of an elongated cone, rounded at the top and pointing downward. The parameters of this cone vary somewhat for different rocks and apparently depend not only on the strength of the rocks, but also on other properties (grain size, type of bond of the grains, presence of fractures, etc.).

Figure 20 shows a diagram of a model hole drilled in a cement block by explosions under water. The curved lines show the configuration of the near-bottom segment and the downward growth of the hole after each ten explosions.

Experiments were also conducted on models with artificially decreased penetration per explosion by introducing materials into the cement block that are very slightly susceptible to explosions, such as resins, and by exploding the charges at some distance from the surface to be shattered. These experiments have shown that a decreased rate of sinking the hole, with constant parameters of charge, gives rise to an increase in the diameter of the hole formed.

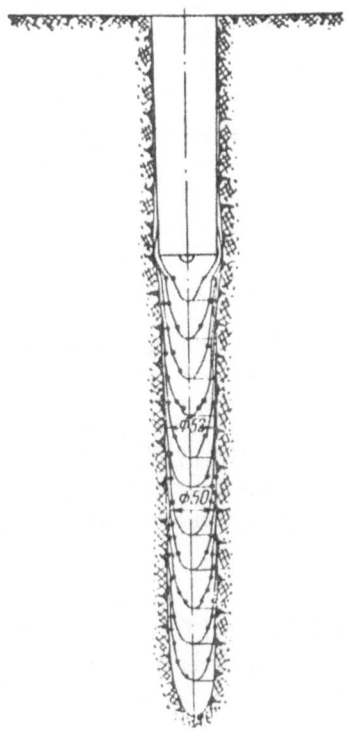

Fig. 20 Formation of a model hole made in a concrete block by explosions under water.

Strength Characteristics of Rocks during Explosive Discharges

The explosive yielding of rocks that have comparatively similar strength characteristics (such as measured on Protod'yakonov's scale) may vary appreciably. Rocks that are strongly plastic (such as clays, gypsum, siltstones, etc.) yield less to explosive deformation than do stronger rocks.

On the basis of experimental data we have obtained and on the basis of theoretical investigations by G. I. Pokrovskii and I. S. Fedorov [39], it may be stated that during an explosion in an infinite solid medium for which the critical tensile (or rupture) strength is much less than the critical compressional strength ($\sigma_{tens} \ll \sigma_{comp}$) shattering in the material, except in a comparatively narrow zone of local effect of the explosion, will be defined by tensional stresses arising in the medium because of lateral thrust. A similar result may be obtained in holes during explosive drilling.

Apparently the critical tensile strength is one of the most important physical-mechanical characteristics that has been established for many rocks under static loading. Under conditions of applied dynamics, especially explosive discharges, the strength characteristics of materials and the behavior of these materials may alter substantially.

During static loading of a solid medium, the chief factor determining shattering of the material is the development of tension within it.

During an explosion the critical strength (σ_{cr}) apparently cannot be considered the only value determining shattering that occurs in the material. This is explained by the fact that the duration of action of the shock wave or dilation wave, producing residual deformations in the medium, is so small (in comparison with the time of deformation) that the impulse of pressure is transmitted to a particular layer of material considerably before the particles in this layer can be appreciably displaced from their positions of equilibrium.

It may be assumed that deformation in the medium will occur by the transmission of a wave impulse I and that it will continue only while the specific impulse in the wave is greater than some limiting value of $I_{0\,cr}$, depending on the susceptibility to explosion of the given material.

The Method of Determining Critical Tensile Strength ($\sigma_{tens\,cr}$) for Solid Media during Explosive Discharges. According to F. A. Baum, during short-period impulses of pressure, solid media such as rocks behave as elastic bodies under tensile stresses considerably greater than the admissible criteria of plasticity obtained from static tests. This fact permits us to consider such questions within the framework of the elasticity theory.

Experiments have been made with rods of Plexiglas. When there is a disturbance due to a small explosive charge at one end of a rod, a shock wave arises in the rod, acquiring the character of a plane elastic compressional wave within a comparatively short distance from the focus of the explosion. As the wave approaches the free end of the rod it is reflected, and a dilational wave is propagated along the rod in the opposite direction. If this wave is sufficiently intense a piece of the rod is broken off. The piece of material is broken off at the instant the tensional stress in the material (as a result of interaction between a dilational wave and the trailing segment of a compressional wave) becomes equal to the dynamic limit of the resistance of the material to rupture.

In order to establish the magnitude of the tensile strength (critical resistance to rupture $\sigma_{tens\ cr}$), it is necessary to know the shape of the elastic wave and the parameters of this wave at the instant it approaches the free end of the rod, the interval of time τ_0 from this instant until the instant the first piece of material breaks off, and also the length l of the piece. The dynamic limit of resistance to rupture is determined in this way as an algebraic sum of the compressional and dilational stresses at the site where the first piece of rod is broken off (Fig. 21).

The shape of an elastic wave, $\sigma = \sigma\ (t)$, may be determined from experimental data on the displacement of the free end of the rod with time:

$$S = f_1(t). \tag{1}$$

By differentiating equation (1) we find, further, the relation of velocity v of movement of the free end of the rod:

$$v = \frac{dS}{d} = f_2(t). \tag{2}$$

The pressure in the compressional wave for any medium is determined by

$$P = |\sigma| = \varrho_0\ ud, \tag{3}$$

where ρ_0 is the density of the medium in the undisturbed state, u is the velocity of displacement of the medium at the wave front, and d is the velocity of propagation of the compressional wave.

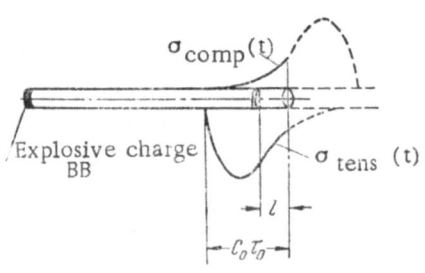

Fig. 21. Position of the trailing segment of a compressional wave and location of dilational wave at the moment the first piece of rod is broken away.

For elastic waves $d \approx c_0$, where c_0 is the velocity of sound in the medium in the undisturbed state. In addition, for compressional waves it may be accepted, with a high degree of accuracy, that

$$u = \frac{v}{2}. \tag{4}$$

Further, by using the relation (4) and the experimental relation (1) we may obtain a graphic relationship that is of interest to us

$$\sigma = f_3(t). \tag{5}$$

In order to determine the nature of movement of the free surface of the rod with time, a method of rapid photographic recording was used, based on a combination of photographic scanning and a microscope. With this device it is possible to magnify the

Fig. 22. Photographic record of displacement s of the free end of rod with time t.

object greatly, since the magnitude of displacement of the free surface of a solid medium, even when acted on by very strong elastic waves, is very small (of the order of tenths and hundredths of a millimeters). A typical photographic record of the moving free end of a rod, obtained by the indicated method, is shown in Fig. 22.

The limiting resistance to rupture for Plexiglas, as determined by the above-described method, has been shown to be, according to Baum, $\sigma_{tens\ cr} \approx 1300\ kg/cm^2$, whereas under static conditions the stress in the same material was found to be considerably less, specifically $\sigma_{tens\ stat} \approx 650\ kg/cm^2$. It may be foreseen that a similar relation is possible for other materials as well, including rocks.

By this method it is possible to determine the critical specific impulse $I_{0\ cr}$ during rupture for the corresponding materials. For Plexiglas the value of $I_{0\ cr}$ has been found to be $5 \cdot 10^{-3}\ kg \cdot sec/cm^2$.

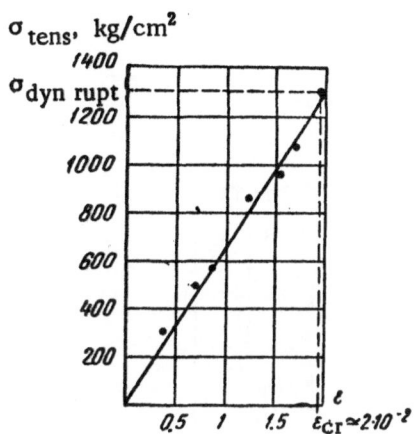

Fig. 23. Relative deformation of a solid medium in relation to the tensile stress σ_{tens}.

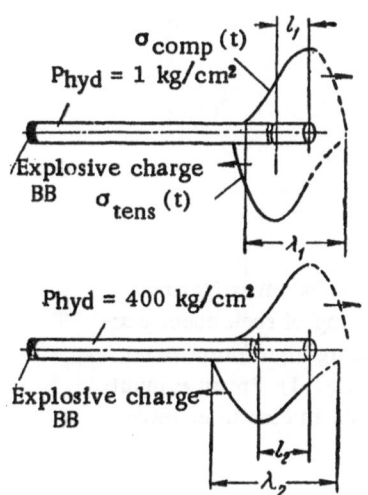

Fig. 24. Position of an elastic wave at the instant the first piece of rod breaks away, at normal and high pressures.

F. A. Baum was also successful in discovering the behavior of Plexiglas during the preshattering interval under the influence of dilational waves. The results of these investigations are shown on a graph of $\varepsilon = \varphi\ (\sigma_{tens})$, where ε is the relative deformation of the material (Fig. 23). In the examined case ε represents the ratio of absolute displacement of s particles in a given segment of rod to the length λ of that part of the rod that has been deformed by a dilational wave. From the graph it may be concluded that Plexiglas, when an explosive charge is applied to it, despite well-defined plastic properties, strictly obeys Hooke's law under uniaxial tension up to the instant of rupture; i.e., it behaves like an ideal brittle body.

The systematic behavior established for Plexiglas during explosive discharges is of a more general character and may be extended, apparently, to most solid rocks.

The Influence of the Surrounding Fluid Medium on the Breaking of Plexiglas during Explosive Discharges. An explanation of the effect of hydrostatic pressure on the intensity with which fragments break off is of interest in studying the shattering of deep-lying rocks by explosions in holes filled with liquid.

The effect of hydrostatic pressure has been tested in experiments with rods on Plexiglas (150 mm long, 10 mm in diameter), the ends of which were subjected to explosive charges (~ 0.1 g) with the rods loaded to produce pressures up to 400 kg/cm². Parallel experiments were conducted in air.

The experiments have made the following facts clear:

1. The intensity of breaking, as determined by the dimensions and number of fragments broken from the rod, substantially increases with deeper burial of the charge in water; this is explained by an increase in the active part of the charge in the presence of an "aqueous envelope."

2. During discharge in a liquid (water) an increase in pressure up to 400 kg/cm² considerably increases the intensity of breaking. For instance, at a pressure of 1 kg/cm² one piece 13 mm in length breaks off a rod, as a rule. At 400 kg/cm² the number of pieces breaking off ranges from one to three, and the average length of the first piece to break off is 26 mm.

3. When the explosions are discharged in air, the change in pressure (up to 400 kg/cm²) exhibits practically no effect on the intensity of breaking. This fact is apparently due to a definite influence of the compressibility of the surrounding medium on the conditions of scattering of the detonation products of an explosive charge.

As already noted, a piece of rod is broken off in that segment where tensional rupture arises through interaction between a reflected dilational wave with the trailing part of an advancing compressional wave (see Fig. 21). However, when there is considerable resistance in the liquid acting on the end of the rod, the movement of particles at the free end of the rod is somewhat retarded, and the velocity of the particles is diminished. In this process the dilational wave will be elongated and its amplitude will decrease. It is clear that under such circumstances critical tensile forces may arise only in segments at greater distances from the free end of the rod, as shown diagrammatically in Fig. 24 for a hydrostatic pressure of 400 kg/cm². It may thus be concluded that on increase of hydrostatic pressure, up to a certain limit, the intensity of shattering in a medium such as solid rock may increase.

Observations during the course of explosive drilling of holes indirectly support the expressed opinions and the experimental data obtained in tests on rods of Plexiglas. During explosions in deep holes under high hydro-

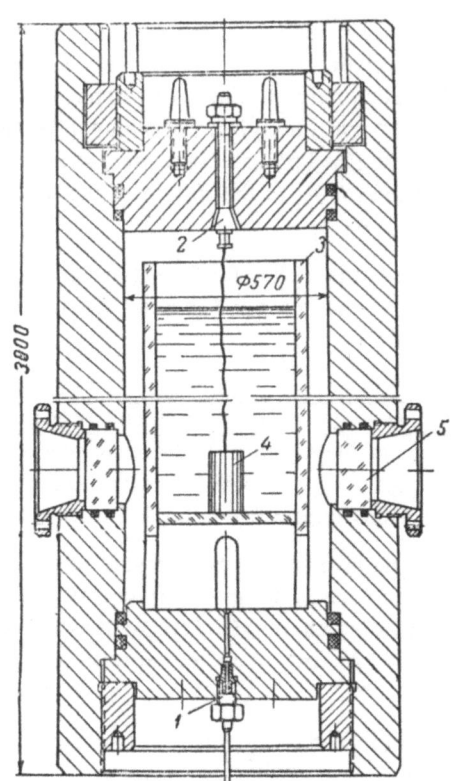

Fig. 25. Device for investigating the properties of explosives and of explosive yielding of rocks at high hydrostatic pressures. 1) Pressure input; 2) electrical lead; 3) iron ring; 4) explosive charge; 5) window for photographic recording.

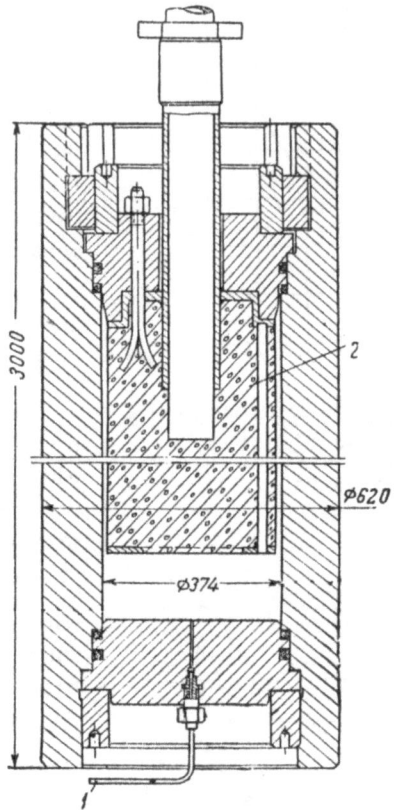

Fig. 26. Device for investigating explosive yielding of rock under conditions simulating high formational and rock pressure. 1) Pressure input; 2) block of rock in elastic envelop.

static pressures, more large fragments of rock were observed to break away from the mass (from the walls of the hole) than from similar rock at shallower depths.

To determine experimentally the strength characteristics of rocks during various forms of deformation produced by detonation of explosive charges, devices were used whose general construction is shown in Figs. 25 and 26. The first of these (Fig. 25) permits the investigation of explosive yielding of various rocks at hydrostatic pressures up to 600 kg/cm² and the determination, by means of a photographic recorder, of the velocity of propagation of shock waves in solid and liquid media and the rate of displacement of the interfaces between different media. The second device (Fig. 26) has been designed for studying the explosive yielding of different rocks under conditions simulating formational and rock pressure, up to 1000 kg/cm^2; in using this instrument the bottom of the model hole may be kept at atmospheric pressure or may be subjected to a liquid pressure as great as 1000 kg/cm^2.

Outline for Forming Holes by Explosions

In the explosive technique of drilling, explosive charges are placed on the surface of the rock under a layer of liquid and detonated at a definite frequency; the dimensions of the charges are small in comparison with the diameter of the cavity formed during the underwater explosions. As a result of the sequential effect of repeated explosions the rock is shattered and the shaft of a hole is formed.

On the basis of explosive drilling in deep holes and also on the basis of a large number of experiments on massive rocks and on models, the following qualitative scheme for forming holes by successive explosions may be outlined [40].

As a result of the action of the detonation products, which produce a very high pressure (near $2 \cdot 10^5$ kg per cm^2) during the initial stage of expansion, the rock at the bottom, in the zone directly beneath the explosive charge, is shattered and separated from the mass. This zone of destruction will hereafter be termed the zone (funnel) of crushing.

Beyond the zone of crushing the effect of the explosion on the rock is found to be inadequate to separate fragments from the mass. However, in a certain zone about the bottom of the hole, varying in depth and called the advance zone of crushing, fractures and other disruptions of continuity develop in the rock. In addition, the explosion produces shock waves in the liquid filling the hole, the effect of which on the walls of the hole leads to the formation of fractures. Furthermore, a certain erosional effect by streams of liquid occurs, leading to movement of the spreading products of the explosion.

The zone where residual deformation of the rock occurs because of transmission of explosive disturbances through the liquid is tentatively termed the zone of hydraulic effect of explosion.

Let us examine the sequential effect of two explosions designated on Fig. 27 by I and II. Let the zone of crushing formed after the first explosion be designated A_1, the advance zone of crushing B_1, and the zone of hydraulic effect C_1. The next charge will be placed at the bottom of the funnel of A_1, and the explosion will produce zones A_2, B_2, and C_2 analogous to the preceding, these later zones partly intersecting the corresponding zones of shattering formed by the preceding explosion.

With repeated explosions, as a result of sequential complication in the zones of crushing, a hole is formed whose diameter, if there are no other zones of shattering, should be nearly that of the aperture of the funnel produced by the repeated action of explosion of an applied charge. The aperture of the funnel, as shown by experiments with charges having the form of bodies of rotation (the ratio of height to greatest diameter ranging from 1 : 1 to 3 : 1), reaches values of two or three times the diameter of the charge in various rocks. However, the presence of advance zones of crushing and of zones of hydraulic effect substantially change the diameter of a hole.

In sections where zones B and C are repeatedly superimposed, the rock is shattered by the formation of new fractures and also by the development of fractures from the effect of earlier explosions. After several explosions these sections of the rock prove to be so shattered that particles are separated from the mass and may be removed from the hole. As a result of repeated explosions new segments of the hole are ever subjected to the effect of the explosions, exposed by the removal of rock particles. The hole is thus enlarged.

With increasing distance from the center of the explosion the pressure at the front of a shock wave in liquid drops and at some distance proves to be insufficient to form or extend fractures. Here the hole is no longer expanded.

In regard to the mechanism of shattering rocks in the various zones of influence of underwater explosions, the following may be said.

During the detonation of a charge on the surface of a rock, a high pressure becomes effective, deforming and shattering the rock to some depth. The pressure of the compressional wave traveling through the rock decreases with distance; consequently, at a certain depth the stress falls below the limit at which shattering of the rock is possible in the process of compression. At greater depths in the rock, during propagation of compressional waves, only individual microruptures and fractures may be developed.

With frequent repetition of explosions, the number and extent of the fractures increase and individual fragments are more easily separated from the rock mass; this becomes possible even under compressional stresses insufficient to shatter the same rock in its initial (unshattered) state.

The indicated explosive shock on the rock is very short-lived, and the pressure of the gaseous products of the explosion falls very rapidly because of the expansion of these products. After the compressional wave there arises a dilational wave, traveling into the depth of the mass [41, 42, 43, 44].

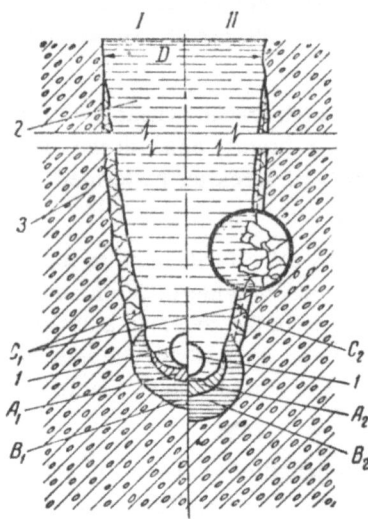

Fig. 27. Schematic diagram of zones of influence on rock of two successive explosions in a hole. I) First explosion; II) second explosion, A_1 and A_2) zones of crushing after first and second explosions respectively; B_1 and B_2) advance zones of crushing, C_1 and C_2) zones of hydraulic effect of the explosions; D) diameter of hole produced. 1) Explosive charge; 2) liquid; 3) rock mass.

The stresses arising in the rock during passage of a dilational wave facilitate further development of cracks. It may be expected that at some depth, where the amplitude of the compressional wave proves to be inadequate to produce residual deformation in the mass of the rock, the dilational wave, even with a smaller amplitude, produces definite shattering. The dilational wave, depending on its parameters at a given point and on the properties of the rock, as with a compressional wave, may cause separation of particles of rock from the mass by a single action or by the superposition of repeated shocks.

The possibility of shattering solid bodies by the passage of dilational waves from several explosions (splitting off) has been demonstrated by the above-described experiments, conducted while subjecting the test samples to hydrostatic pressures up to 400 kg/cm^2.

Thus, the shattering of rocks within the zones of crushing and the advance zones of crushing is determined by the mechanical effect of compressed products of explosion of greatest density.

In the zone of hydraulic effect, the magnitude of the pressure during decline of a shock wave is small in comparison with the magnitude in the zone of crushing. On the basis of approximate evaluation of the parameters of a diminishing wave, it may be stated that a compressional wave traveling through the rock in the zone of hydraulic effect of the explosion cannot produce any significant shattering in the entire mass. The formation of fractures in this zone occurs chiefly because of repeated development of positive and negative stresses and because of the wedging effect of the liquid filling the fractures in the rock.

The differences in mechanism of shattering in the three zones of influence of the explosion define different degrees of comminution of the rock particles. In the zone of crushing the rock separates from the mass in finer particles than in the advance zone of crushing or in the zone of hydraulic effect.

The discussed outline for forming holes by explosions is in keeping with experimental data.

It has been confirmed experimentally that the zone about the bottom of the hole has a form resembling a truncated cone with a diameter of the small base equal to three or four times the diameter of the charges. The large base of the cone, the diameter of which is equal to the diameter of the ultimate hole, lies at a distance from the lowermost point of the bottom (for solid rocks) six or seven times the diameter of the hole.

When the effectiveness of the explosions is low and the penetration is "null," the explosive disturbances traveling through the liquid repeatedly act on a particular segment of the hole, producing shattering in the walls of the hole and forming cavities. The growth of these cavities in radial and axial directions slows and then stops when the limiting dimensions are reached, depending on the strength of the explosions and on the yielding capacity of the rocks.

When the explosions are set off in water, the diameter of the hole produced is three or four times that of a hole free of liquid and produced by explosions under identical conditions with similar explosive charges.

The nature of the shattering of solid materials in the zone of hydraulic effect of the explosion was studied by F. A. Baum and E. B. Kagan by detonating small charges in blocks of Plexiglas and cement. Figures 28 and 29 show the development of cracks in a block of Plexiglas after the first and second explosions respectively.

An examination of the experimental data permits us to distinguish schematically three types of surfaces of fracture development in solid media (Fig. 30):

1) The plane R passing through the axis of the hole, forming fractures r on the lateral surface of the hole parallel to the axis (radial fractures);

2) The plane G perpendicular to the axis of the hole, forming circular fractures g, disposed in series one above the other on the lateral surface of the hole;

3) Concentric surfaces C coaxial with the hole (weakly developed), forming fractures c in the mass of the medium (beyond the hole), which appear on the horizontal free surface as concentric fractures.

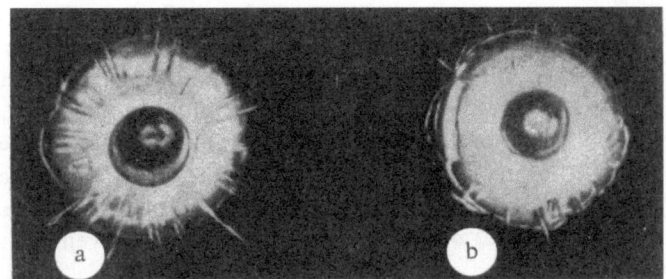

Fig. 28. Development of fractures in a block of Plexiglas after the first explosion. a) View from above; b) view from below.

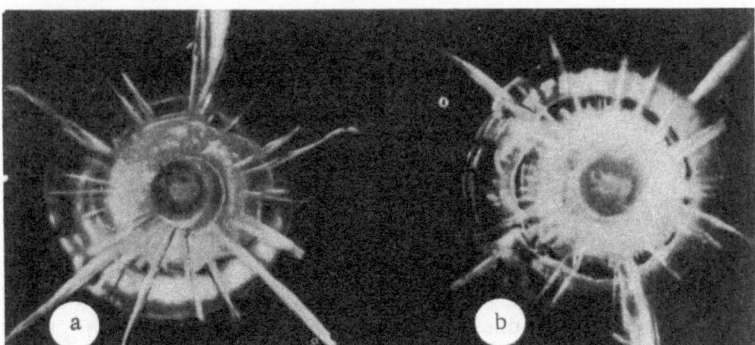

Fig. 29. Development of fractures in a block of Plexiglas after the second explosion. a) View from above; b) view from below.

The intersection of the radial and circular fractures forms a network on the lateral surface of the hole. With repeated action of shock waves, cracks are developed, and in the zone of hydraulic effect relatively large fragments of crushed material are broken off; the dimensions of these fragments are comparable with the dimensions of the charge, and in some cases exceed them considerably.

It was noted above that the effect of the explosion in the hydraulic zone is characterized by separation of the largest fragments from the mass of solid material, the dimensions of which are directly related to the system of surface fracture development. For the system of fractures parallel to the axis of the hole, the dimensions of these fragments are defined by the size of the angle between adjoining planes of fracture growth and the depth to which the fractures penetrate the mass.

For the circular fractures the parameters determining the dimensions of the fragments breaking away from the solid medium are the distance between the planes of these fractures perpendicular to the axis of the hole and the depth to which the fractures penetrate the rock. For the concentric fractures, rather weakly developed (at least in the indicated experiments on Plexiglas), the principal parameters determining the size of the detached fragments apparently must be the "concentration" and the depth of occurrence within the mass of the fracture surfaces intersecting the surface of the hole (the planes passing through the axis of the hole and those perpendicular to the axis).

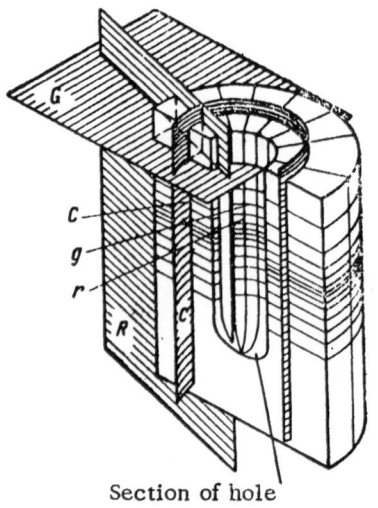

Section of hole

Fig. 30. Schematic section of a hole and of surface fracture growth in a solid medium during underwater explosions.

The Principal Parameters of an Underwater Explosion and Its Zone of Influence

Expansion of the Gas Bubble Containing the Products of Explosion. When an explosion occurs on the bottom of a flooded hole, the products of explosion form a gas bubble which, if the hole is sufficiently deep, completes several pulsations (successive expansions and compressions) before it floats to the surface. We shall confine our examination to some parameters for the case of a spherical explosion, for the first pulsation, which possesses greater energy than the succeeding pulsations. These parameters include, above all, the pressure P, the volume of the products of explosion, and the time required for the expansion of these products. For a spherical charge the relation between instantaneous pressure and volume (or the instantaneous radius r of the gaseous sphere) of the products of explosion is given in the form

$$P = P_0 \left(\frac{r_0}{r} \right)^m = P_0 \left(\frac{v_0}{v} \right)^{\frac{m}{3}}, \tag{6}$$

where P_0 is the average pressure of the products of explosion shortly after detonation of the charge (in this case taken to be immediately), r_0 the radius of the spherical charge, m a constant in the interval of which r varies, and v and v_0 are the instantaneous and initial volumes of the products of erosion respectively.

Figure 31 shows this relation graphically.

By using the empirical relation $t = f(r)$ relative to the first pulsation of a gas bubble from an underwater explosion and also using (6) it is possible to obtain a relation between pressure of the products of erosion and the time required for their expansion.

In computing the hydrostatic pressure P_h it should be borne in mind that the effect of this factor on the expansion of the gas bubble is very marked (in the late stages of expansion). According to R. Cole [45] the maximum radius of the gas bubble r_{max} is inversely proportional to $P_h^{1/3}$, and the time of expansion of the bubble from its initial radius r_0 to r_{max} is inversely proportional to $P_h^{5/6}$. This phenomenon clearly points to the intensity of streams of liquid displaced by gases during the explosion.

Parameters of the Shock Wave Propagated Through the Liquid. Besides the direct effect of the products of explosion, during their expansion, on the bottom of the hole, a very important role in shattering rock is played by the shock waves propagated through the liquid during an explosion.

The maximum pressure at the front of a shock wave in water immediately next to the focus of the explosion reaches 200,000 kg/cm² for liquid explosives. However, for a spherical explosion P_{max} falls off with distance from the focus of explosion, very rapidly at first and then more slowly. The relation between P_{max} for these explosives and the dimensionless distance r_1/r_0 of a point in the liquid from the center of the explosion is shown graphically in Fig. 32, from data of K. P. Stanyukovich and R. Cole (r_1 is the distance from the point in the liquid to the center of the explosion).

Spherical propagation of a shock wave occurs only till the wave reaches the walls of the hole; then, with increasing distance from the focus of explosion, the shock wave changes into a plane wave. This leads to a comparatively slow diminution in P_{max} of the shock wave and a marked destructive effect of explosive disturbances transmitted by water to the walls of the hole, to considerable distances, to the zone of hydraulic effect of the explosion.

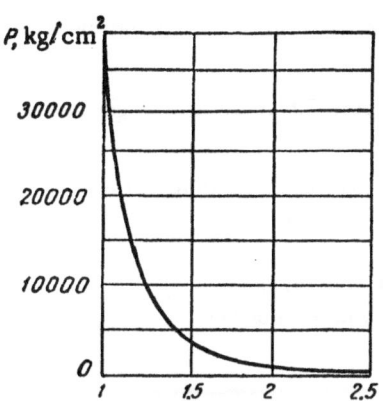

Fig. 31. Relation of pressure P of the products of erosion to the relative value r/r_0 of the radius of the gas bubble (first pulsation).

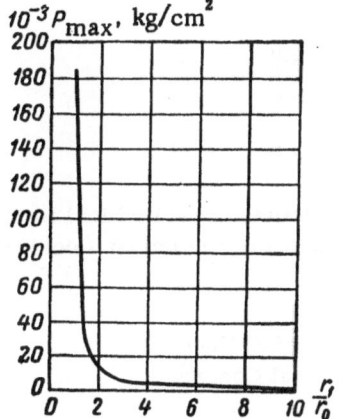

Fig. 32. Approximate relationship between maximum pressure P_{max} of a shock wave at a point in the liquid and the dimensionless distance r_1/r_0.

The Zone of Crushing and the Zone of Hydraulic Effect of Explosion. The shattering of rocks in the zone of crushing depends chiefly on the direct effect of the products of explosion, which are at a high pressure. Because of the rapid drop in pressure during expansion of the products of explosion, the crushing effect of the explosion ceases shortly after detonation of the charge; this fact also explains the comparatively small dimensions of this zone (for explosive charges of about 30 ml the experimentally obtained diameter of the zone of crushing in solid rocks with a compressive strength of 800-2000 kg/cm^2 is 70-90 mm).

The greatest depth of the zone of crushing is associated with the intensity of the effect of the products of explosion, which in turn is determined by the size and duration of the action of the pressure on the various sections of the material being crushed. Naturally, when the form of the charge is symmetrical with the axis of the hole and the initiator is also symmetrical, the segments of rock beneath the central part of the charge will be exposed to the effect of the products of explosion at the very beginning of the process of expansion; i.e, the magnitude and duration of the action of the pressure of the products of explosion will be greatest for these segments. Therefore, the depth of the zone of shattering, reaching a maximum in the indicated segments and decreasing from the center toward the periphery because of the diminution in intensity of the explosive disturbances, becomes zero.

It has been discovered that the dimensions of the zone of crushing depend on the form of the charge, the intensity of the initiating element, and the distance between the initiator and the charge (discussed in Chapter 4).

The dimensions of the zone of hydraulic effect of the explosion depend on the intensity of the shock waves and of the streams generated in the liquid during the explosion; they also depend on the properties of the rock.

According to experimental data the upper boundary of the zone hydraulic effect is a considerable distance from the bottom of the hole. In an actual hole (crushing strength of the rock about 800 kg/cm^2) with a diameter of 300 mm, during successive explosions of charges about 40 g is weight, the zone of hydraulic effect extends for about 2 m. This supports the previously expressed view concerning the comparatively slow drop in intensity of shock waves during their propagation in the hole.

It is clear that the size of charge is one of the principal factors defining the parameters both of the zone of crushing and of the zone of hydraulic effect of explosion.

With some schematization of the complex phenomena that appear during explosive drilling of a hole, it is possible to express a relationship between the dimensions of the hole and the size of the charge [40]. Thus, after n successive explosions in the drilling of a hole, the following relationship holds:

$$H = nAG^k, \qquad (7)$$

where H is the depth increment of the hole in m, A is a coefficient depending on the properties of the explosive used, G is the weight of the charge in kilograms, and k is an exponent, equal to $^1/_3$.

In experiments with model charges of similar form on various rocks, values of A ranging from 0.035 to 0.050 were obtained.

The diameter of the hole formed, in relation to the weight of the charge, is

$$D^2 h_1 = BG^m, \tag{8}$$

where $h_1 = H/n$, B ranges from $9 \cdot 10^{-3}$ to $14.5 \cdot 10^{-3}$, and m is an exponent near unity.

Below we show the results of investigations on models, the effect of size and form of charge on the average penetration and diameter of the hole.

In changing the size of charge from 0.3 to 3.0 ml or, correspondingly, from 0.5 to 5.0 g, the geometric form was approximately preserved.

Experiments on sinking model holes with charges of various sizes were conducted on identical concrete blocks buried in 6 m of water. The results of these experiments are shown in Table 4.

TABLE 4. Results of Explosive Drilling of Model Holes in Concrete (under a Layer of Water) by Explosive Charges of Various Sizes of Approximately the Same Form

Charge of liquid explosive, ml	Average penetration per explosion, mm	Ultimate Diameter of model hole, mm	Average volume of concrete shattered by a single explosion, ml
0.3	2.5	42.0	3.5
0.6	4.2	43.0	6.0
1.2	6.4	50.0	12.5
3.0	9.4	68.0	35.0

The influence of the geometric parameters of a charge (1.5 g of liquid explosive) of cyclindrical form, with the igniter at the top of the charge, on the effectiveness of the explosion (in air) was tested on blocks of granite.

TABLE 5. Comparison of the Effectiveness of Explosions (in Air) During Drilling in Model Holes in Granite for Various Cylindrical Charges (weight of 1.5 g)

Dimensions of charge			Average penetration per explosion	
diameter d, mm	height h, mm	$h : d$	mm	%
16.0	5.0	0.32	3.2	64
10.9	10.9	1.00	5.0	100
8.6	17.2	2.00	7.0	140
7.5	22.5	3.00	5.6	112

Table 5 shows the results of experiments with cylindrical charges, for which the ratio of height h to diameter d ranged from 0.32 to 3.0. It may be seen from the table that the effectiveness of the charges of cylindrical form increased with relative elongation to h : d = 2.

It should also be noted that the author established the relationship between penetration per explosion and the position of the point of initiation in the charge. For a model charge with the ratio h : d = 2, transfer of the point of initiation from the upper to the lower end of the charge decreased the penetration per explosion by a factor of 2.5 (according to data of A. I. Gol'binder and E. B. Kagan).

The Effect of Hydrostatic and Rock Pressure on the Shattering of Rocks by Explosions

In holes at depths greater than 1500-2000 m, a decrease in penetration rate is observed in comparison with the rate at depths down to 1000 m, a fact for which it may be assumed that there are several constant factors influencing the effectiveness of the explosions: technological complications, changes in the physical-mechanical properties of the rock because of hydrostatic and rock pressure, effect of hydrostatic pressure on the action of the charge and on the properties of the explosive, and effect of hydrostatic and rock pressure on the shattering of rocks by explosions.

It is very difficult to distinguish the last factor by experiments in actual holes, since experimental data have shown that two of the first factors act simultaneously at great depths. Therefore, besides investigations on technological factors (Chapters 8 and 9), on the properties of rocks under high hydrostatic, formational, and rock pressure (Chapter 3), and of the properties of the explosive charge and action of the charge under pressure (Chapter 4), the effect of hydrostatic and rock pressure on the shattering of rocks by explosions was also studied.

Hydrostatic Pressure. During the initial stage of the explosion, the pressure of the gases is measured in tens of thousands of atmospheres; consequently the counter pressure (hydrostatic pressure), not exceeding several hundred kilograms per square centimeter even in the deepest holes, has practically no effect at this stage of the explosion on the relationship between radius of the gas bubble and the time required for its expansion.

Experimental data cited by R. Cole [45] confirm the identity of the initial segments of the curves $r = f(t)$ at various hydrostatic pressures; this fact requires also the identity of the curves $P = P(t)$ defining the dimensions of the zone of crushing. From this it may be concluded that the depth at which shattering of rock from compressive stresses extends into the zone of crushing is practically independent of the hydrostatic pressure under conditions in which the explosive yielding of the rocks is preserved.

However, the supposed mechanism of shattering rock in the zone of crushing is also related to tensile stresses arising when the explosive forces are relaxed. To the extent that hydrostatic pressure in some measure retards the displacement of elements of the rock (toward the liquid) at the interface of the media, it should be considered together with the possibility that some weakening of local action of the explosion during a marked increase in hydrostatic pressure may occur.

Investigations on models of the effect of hydrostatic pressure on the effectiveness of explosions [16] have been made on blocks of concrete (compressive strength of 320-400 kg/cm^2) at various heights of the water column above the charge (KD-8). Similar experiments were made on blocks of granite (compressive strength of 1700-2000 kg/cm^2) embedded in concrete in a steel ring. With a layer of water up to 300 mm deep the block of granite remained at the surface. In experiments with higher hydrostatic pressures the block was submerged to a proper depth in a hole filled with water. At the same time these experiments made clear the effect of the height of the column of liquid above the charge on the dimensions of the model hole formed.

The results of the observations are shown in Table 6, in which the average data for 50-250 parallel experiments are given. The results of the experiments, even under laboratory conditions, are not distinguished by good reproducibility. Nevertheless, the very large number of experiments performed permit us to conclude that the sinking of a model hole by explosion, i.e., the local effect, reaches a maximum value when there is no liquid medium possessing high inertia and partly suppressing the dilational wave, which during explosions in air permits more intense shattering of rocks in the shatter zone of explosive action.

A substantial decrease in the effectiveness of an explosion begins with a comparatively thin layer of liquid over the bottom of the hole, equal in thickness to 5-10 times the diameter of the hole. A further increase in the column of liquid to 800 m has no marked influence on the effectiveness of the explosions.

TABLE 6. Effect of Height of Column of Liquid above an Explosive Charge on the Effectiveness of Explosions and the Diameter of Model Holes

Material in the model	Ht. of column of water above the charge, m	Aver. penetration per explosion, mm	Aver. volume of material crushed in a single explosion, ml	Ultimate diameter of the model hole, mm
Concrete	0	4.3	3.1	29
	0.025	3.5	15.5	75
	6.0	3.1	6.6	52
	55.0	2.8	7.3	57
	800	3.0	7.9	58
Granite	0	3.5	1.9	29
	6.0	3.0	5.9	50

Rock Pressure. Most deep-lying rocks may change their physical-mechanical properties and explosive susceptibility through compaction by high rock (and hydrostatic) pressure. The effect of rock pressure on the shattering of solid media such as rock by explosions has been investigated in model holes on concrete and cement blocks of approximately identical strength (compressive strength of 180-200 kg/cm^2).

The device for studying this factor consisted of a high-pressure vessel, in which a block of test material with a model hole was placed. A rubber sack was placed on the outer surface of the block. Stress in the test material corresponding to a given rock pressure was created by pumping fluid between the rubber sack and the inner walls of the vessel. Atmospheric pressure was maintained constantly at the bottom of the model hole.

TABLE 7. The Effect of Rock Pressure on the Effectiveness of Explosions and on the Diameter of the Model Hole

Material in the model (block)	Pressure on the block, kg/cm^2	Simulated depth of hole, m	No. of experiments	Average penetration per explosion, mm	Ultimate diameter of model hole, mm
Concrete	50	185	70	0.94	28—29
	50	185	60	0.90	30—31
	500—504	1850—2000	20	0.70	—
	550	2040	70	0.69	32
	630	2330	70	0.60	34
Cement	50	185	30	0.84	30—31
	630	2330	30	0.50	32—34

The model charge was 0.1 ml of liquid explosive mixture.

The adopted scheme of modeling approximately satisfied the conditions of simulating, with a small block, a monolith of rock in the bottom zone of a deep hole. According to computed data, when the depth of the model hole is more than three times the diameter, the stresses on the ends and lateral surface of the block differ from the stress in an undisturbed mass of rock by 10%.

It was assumed that rock pressure varied with depth according to the law $P_{rock} = \gamma H$, where H is the depth of the hole and $\gamma = 2.7$ g/cm^3, the average density of the rock.

The data from laboratory tests (Table 7) show that the explosive susceptibility of the model material depends on rock pressure. An increase in pressure corresponding to a change of depth on the order of 200 m at a depth on the order to 2000 m leads to a decrease in penetration of 25-35% for concrete and 40% for cement. From this it follows that the effect of rock pressure on the explosive susceptibility may vary for different rocks.

The results of model tests agree qualitatively with observations made during explosive drilling of holes of various depths and confirm the fact that a change in the properties of the rock due to rock pressure (see Chapter 3) leads to worsening of drilling characteristics for drilling either by mechanical instruments or by explosions.

Chapter 3

THE FORMATION OF HOLES BY EXPLOSIONS IN ROCKS
IN THEIR NATURAL MODE OF OCCURRENCE*

The results of investigating in models the formation of holes by explosions (Chapter 2 and [16]) were checked by explosive drilling of holes in rocks in their natural occurrence.

In Chapter 3 we examine the data from observations on the effect of individual explosions on rock surfaces and on the process of drilling holes 2-3 m deep by explosions; general information is reported on the changes in physical-mechanical properties of the rock subjected to hydrostatic and rock pressure, and some systematic relations obtaining during explosive drilling of holes down to depths of 2800 m [17] are discussed.

Drilling Shallow Holes

In deep holes the charges must be exploded on the bottom under a column of liquid (water, clay mud) reaching a height of several thousand meters. To reproduce such conditions in experiments conducted on the surface is impossible. It has therefore been necessary to select an amount of liquid above the charge (the value of liquid "tamping") acceptable for the experiment and, together with this, sufficient to secure, so far as possible, the full effect of the explosion.

From the theory and practice of blasting it is well known that, when using superposed explosive charges, the principal factor is not the strength, but the mass of the "tamping." An increase in tamping above some limit (in order of magnitude the optimum length of "tamping" approaches the size of the charge) does not increase the effect of the explosion.

The conditions for exploding comparatively small explosive charges (up to 50 g) in shallow and deep holes are essentially different. In a deep hole there is no ejection of liquid; during explosions in a shallow hole a considerable part of the liquid is ejected with high velocity,**producing supplementary destructional effects on the walls of the hole.

The surface layers of rock generally possess marked structural inhomogeneities, intersecting fractures, weaknesses from weathering, and so forth. Extensive spalling during an explosion is due, furthermore, to the nearness of the free surface. Therefore, in sinking shallow holes in a quarry it is difficult to attain good reproducibility of results in parallel experiments.

---

* Translator's note: Two terms, extensively used in this chapter, may be clarified here. These terms are: effectiveness of explosions (or blasting) and explosive yielding (or susceptibility to blasting). The first signifies the penetration per explosion (increment of hole), referring to the amount of rock crushed and ready for removal by some flushing action. The second refers to the potential a rock has, for a given explosive charge. It may also be measured by penetration rate, the rock permitting the greater penetration, for a given charge and uniform conditions of flushing, having the greater explosive yielding (or susceptibility to blasting).
** Up to 100 m/sec for the adopted charges of liquid explosive of 33 ml (50 g).

Experiments with individual explosions on rock surfaces and in sinking shallow holes were performed on limestones, granites, and clays with variously sized elongated cylindrical charges (h : d = 2 : 1) of liquid explosive mixture.

The size of explosive charge was determined from conditions obtaining in various rocks in holes 200-300 mm in diameter. The charges were placed in the central part of the hole bottom and were exploded with the hole filled to various heights with clay mud (viscosity of 25-30 sec according to SPV— funnel test, and specific gravity of 1.15-1.18) or water.

After each explosion the crushed rock was cleaned from the hole.

Simultaneous observations were made on the size of rock particles broken from the bottom and walls of the hole during an explosion, and studies were made on the effect of cleaning away the crushed rock on the effectiveness of the explosions and on the diameter of the hole.

Experiments in Limestones. The relationship between size and configuration of the hole, on the one hand, and the magnitude of explosive charge (10, 15, 20, and 30 ml) and mass of liquid "tamping" above the charge, on the other, was tested in natural outcrops of dolomitized limestones having a crushing strength (uniaxial) of 400 kg/cm^2 (Testovskii and Shchurovskii stone quarries in the Moscow district).

From an examination of Table 8, in which the results of some of the experiments are shown, it may be seen that explosions under a layer of clay mud 10-30 cm thick, which had been merely poured on the bottom over explosive charges of 10, 15, and 20 ml, form holes (1, 2, 3, 5, and 9) with diameters ranging from 270 to 320 mm. Explosions of charges of 30 ml increase the diameter of the hole to 360-370 mm (hole 4).

The shattering effect of streams of liquid ejected from the hole by the explosions is most strongly expressed in surperficial cracks, less in strong layers of rock.

To examine the erosional effect of liquid on the walls of a hole, explosions were also discharged with smaller amounts of clay mud above the charge (under a layer 10 cm thick). In this experiment, for a charge of 20 ml, the diameter of the hole was reduced to 230 mm (9), as against 330 mm when the charges were set off under a layer of clay mud 30 cm thick. By placing on the bottom of the hole (6) a metallic model of one of the possible types of instruments for explosive drilling, the erosive effect of streams of liquid on the walls of the hole increased markedly and the diameter was enlarged to 370 mm, as compared with 330 mm in experiments without the model instrument (explosive charge of 20 ml).

Cleaning shattered rock from the bottom proves to have a great influence on the effectiveness of the explosions. For example, in sinking hole 5 without cleaning the bottom, the effectiveness of the explosions was approximately but half that for the other holes.

The size of the rock particles breaking off in the shatter and hydraulic zones of explosive activity substantially influences the cleaning of the bottom and the removal of the particles from the hole. It has been observed during experiments that individual large pieces of rock (up to 200 g) are torn off the upper layers. The breaking away of such pieces ceases when the hole deepens to approximately one meter. Without considering these pieces, individual fragments do not exceed a dimension of 20 mm, according to sieve analysis. The content of particles larger than 10 mm is 20%, 3 to 10 mm 70%, and less than 3 mm 10%.

Experiments on Granite. Because of considerable inhomogeneity in limestones, further experiments were made on natural exposures of granite at the Sokolovskii deposit (Zhitomir district), which has comparatively uniform properties (specific gravity of 2.68, porosity of 0.6%, crushing strength of 2000 kg/cm^2, tensile strength of 120 kg/cm^2, and hardness of 94 on Shore's scale).

A funnel is formed when an explosive charge is detonated on the surface of the granite. After the rock crushed by the explosion is removed, one may observe a crust-like layer of partly shattered granite, easily detached from the parental mass. This layer is intersected by numerous fractures; on individual pieces one may see transitional zones, exhibiting different degrees of shattering by the explosion. The surface of such pieces facing into the funnel consists of finely granulated crystals of the granite, separating under the pressure of the fingers. The surface nearest the granite mass is formed chiefly of unbroken crystals and is crossed only by individual fractures.

TABLE 8. Results of Experiments on Drilling Shallow Holes by Explosions in Limestones (crushing strength of 400 kg/cm²) (Testovskii and Shchurovskii stone quarries, Moscow District)

Hole No.	Size of charge, ml	Layer of liquid above charge, cm	Average penetration per explosion	Ultimate diameter of hole, mm	Remarks
1	10	30 (clay mud)	6.6	310 × 320	—
2	15	The same	10.5	280	—
3	20	" "	12.5	330	—
4	30	" "	14.6	360 × 370	—
5	15	" "	6.1	270	Bottom not cleaned.
6	20	" "	10.8	370	Model instrument for explosive drilling, 240 mm in diameter, placed above bottom of hole.
7	15	30 (water	11.9	400	—
8	15	Hole filled to surface (water)	14.2	200 × 450	—
9	20	10 (clay mud)	13.4	230	—

Fig. 33. Fragments of granite "crust," broken off in the shatter zone of explosive action.

Figure 33 is a photograph of fragments detached from the crust of partly shattered granite, and Fig. 34 shows thin sections of the granite: one of rock unmodified by explosion, the other of the "crust" (magnified 16 times).

As a rule the depth of the shatter funnel during explosions in air is greater than that during explosions under a layer of water; this relationship is in agreement with the experiments on models (see Table 6).

The detonation of charges on the surface of the granite or in a shallow hole (water free) produces spalling of comparatively large pieces in the surface layers; the size of these pieces increases when the hole is filled with water. Because of this part of the hole at the mouth, in holes formed by explosions, has a conical shape.

From a comparison of the experimental results with various heights of the water "tamping" and with different depths of hole, one may clearly see the effect of shock waves propagated through the liquid (during an explosion) in shattering the walls of the hole. In hole 3, for instance, which was drilled to a depth of 1.3 m with a water layer of only 15-25 cm (Fig. 35),* the shattering of the rock was restricted chiefly to the zone moistened by water. On further deepening of the hole, the hole was filled to the brim with water before each explosion; this water caused considerable erosion in the upper part when it was ejected by the force of the explosion. When the hole reached a depth of 1.7-1.8 m, further widening ceased and the diameter remained approximately constant.

It is known from the literature [15] and has been confirmed by experiments on models (Table 3) that the depth of the zone of local explosive effect (funnel of crushing) depends but little on the strength of the solid

---

* The outlines represent position of the walls of the hole after each ten explosions.

medium being crushed. Therefore, if a layer of shattered rock remains on the bottom of the hole because of improper cleaning, the depth of the funnel of crushing for the next explosion is diminished in height by practically the thickness of this layer.

As in all the preceding observations (experiments on models and in sinking holes in limestones), the sinking of holes in granite has confirmed the fact that cleaning crushed rocks from the bottom has a substantial influence on the depth increment of the hole formed by explosions and on the diameter of the hole. When the effectiveness of the explosions is diminished, one always observes a rapid increase in the diameter of the hole.

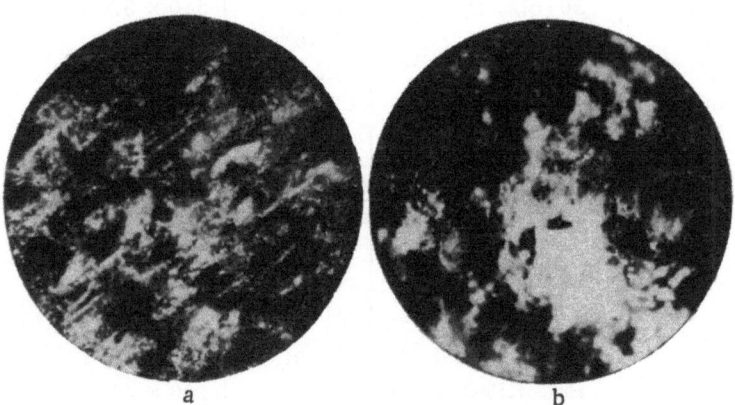

a   b

Fig. 34. Thin sections of granite. a) Unmodified by explosion;
b) "crust."

The effect of the size of charge on the effectiveness of the explosions in granite was tested with 15 and 26 ml of liquid explosive mixture. The ratio of average effectiveness obtained for the first and second charges is $9.6/8.1 = 1.18$.

In keeping with the theory and practice of blasting, the radius of the local explosive effect of the charge is assumed to be proportional to the cube root of the weight of the charge. In the given example the computed ratio of radii of effect for the two tests charges, $\sqrt[3]{26/15} = 1.2$, almost the same as the experimental result, cannot be expected to indicate the degree of precision to be obtained in all experiments, because of the lack of uniformity of properties within a single rock and because of the probability of substantial variations in conditions of conducting parallel experiments (such as in differences in cleaning the bottom of the hole).

The rock extracted during cleaning of the hole was collected, and individual samples from each experiment were dried and analyzed for size distribution. The results obtained are indicated in the following table:

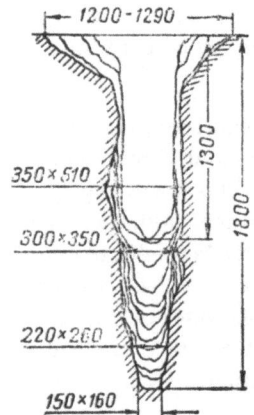

Fig. 35. Diagram of hole development in graphite with various depths of water above the explosive charge.

Quantity of granite detached by detonation of a single
    charge (40 g), kg   2.22

Content of particles (in %) of size (in mm):

greater than 12	38
7-12	10
5-7	6
3-5	7
0.5-3	17
smaller than 0.5	22

The fraction of particles larger than 12 mm contain from one to six comparatively large spalls of granite in eacn sample (the largest range up to 40 mm).

Experiments in Clay. Experiments in sinking shallow holes by exploding superincumbent charges were also made in a bed of clay about 2.5 m thick, in a quarry of the Moscow district. The explosions were started on the bottom of a previously prepared hole about 200 mm in diameter and 0.5 m deep. As in solid rocks, only in greater degree, the dimensions of the hole formed in clay depend on the amount of liquid above the explosive charge.

In clay the increment of depth in the hole for a single explosive charge of 20 ml (30 g) is 8-10 times that when sinking holes, with the same charge, in limestone, and the diameter is 1.3-1.5 times the latter. In surface clays the required diameter of the hole (200-300 mm) may be attained by decreasing the charge to 10 ml (15 g); when this is done the effectiveness of the explosion is reduced almost 30%.

The hole drilled by explosions in clay is cylindrical, the diameter varying but insignificantly throughout the entire hole; only in the very uppermost part is any extensive cone-shaped enlargement noted.

The work of deforming and shattering a medium by explosions, other conditions remaining the same, depends on the density and compressibility of the medium.

An approximate calculation has shown that, when an explosion occurs at the interface between two media—water and rock, considerably less energy is expended on deformation of the rock than on compression and displacement of the water because of the greater density and lower compressibility of the solid medium. However, the compressibility of water is low, and water therefore transmits this energy (shock waves, currents) with comparatively small losses to the walls of the hole, producing shattering of the rock at considerable distances from the focus of the explosion. In this case, as was shown earlier, the ultimate diameter of the hole is substantially larger than the diameter of the funnel of crushing. When air is the medium above the charge, most of the energy is transmitted by the air; but in this circumstance, because of the great compressibility of the air, there occurs an irreversible loss of energy, and, consequently, the effect of the explosion on the walls of the hole will be substantially weaker than when the transmitting medium is liquid. This conclusion is supported by the fact that the diameter of a hole drilled without water is considerably smaller than when water fills the hole.

In order to determine the effect of the medium surrounding the explosive charge (40 g) on the action of the explosion, shallow holes were drilled by explosions under water and without any liquid.

After several explosions in air a funnel 60-70 mm deep and 300-350 mm in diameter at the top was formed on the surface of granite. During the subsequent penetration the surface was not shattered and a hole of true cylindrical form was developed, 80-90 mm in diameter (as against 300-350 mm for underwater explosions). The average penetration per explosion in granite was 16 mm in a hole with no water. In producing this hole the size distribution of crushed materials differed from the distribution in water-filled holes (formed by explosions) in the predominance of strongly comminuted particles; the following table shows the distribution:

Quantity of granite detached by detonation of a single charge, kg	0.22

Content of particles (in %) of size (in mm):

greater than 7	7
from 5 to 7	2
from 3 to 5	9
from 0.5 to 3	20
smaller than 0.5	62

The particles of granite extracted after an explosion were crossed by numerous fractures; they consisted of clumps of crushed crystals. Structurally, altered crusts of granite, predominant features of underwater explosions when sinking holes by explosions, were not observed after explosions in air.

Thus, experiments on sinking holes in solid rocks have shown that, in keeping with the results of investigations on models, the penetration per explosion reaches a maximum when no liquid is present over the explosive charge. When liquid is absent the volume of rock crushed by an explosion is but one-tenth to one-twelfth the volume obtained when explosions of similar charges are made under a layer of water.

Experimental investigations [18, 46, 47, 48] have established the fact that many rocks become denser and stronger with depth and, consequently, have correspondingly poorer drilling characteristics. Observations in practice have shown that, beginning with a certain (considerable) depth, further decrease in the rate of ordinary mechanical drilling ceases, and the rate is stabilized at a very low level of efficiency.

American investigators have reported on laboratory tests on the relationship between penetration rate and hydrostatic pressure [46, 47]. On the basis of comparisons of data from tests at pressures up to 350 kg/cm^2 and from observations of commercial operations they have made the following conclusions:

1) Decrease in penetration rate is a consequence of increase in hydrostatic pressure; there is a maximum pressure above which the drilling rate practically ceases to diminish, and this maximum pressure is different for each rock;

2) With increase in hydrostatic pressure the decrease in drilling rate is greater in easily drilled soft rocks than in strong rocks.

Hydrostatic pressure does not affect the drilling rate in all rocks. The rate in permeable rocks (such as sandstone) does not depend on hydrostatic pressure. Deep-lying beds of sandstone under high hydrostatic pressure drill as rapidly as similar beds near the surface at low hydrostatic pressure.

Most other rock formations show greater difficulty in drilling, since the pressure of the flushing liquid begins to be effective only on such rocks. Thus, the drilling rate in some shales is cut in half with comparatively low hydrostatic pressures (beginning with 35 kg/cm^2).

The use of air and gas as a flushing agent considerably increases the rate of drilling. In Pennsylvania two holes 2100 m deep were drilled under similar conditions, one with clay mud in 510 hr and the other with air flushing in 240 hr, by mechanical methods.

The cited data on relationship between permeability of rock and effect of hydrostatic pressure on drilling rate indicate that the change in physical-mechanical properties of the rock (density, resistance to crushing) is a function of compression of the rock through decrease in volume of pore space because of hydrostatic pressure.

The maximum hydrostatic pressure, above which there occurs no further deterioration of drilling characteristics, corresponds to the cessation of compaction of a given rock because of the limiting diminution (for a given pressure) in volume of pore space or because of the permeability that appears at a definite value of hydrostatic pressure. In this case, when the rock is subjected to an explosion, the hydrostatic pressure may suppress the dilational wave, as has been pointed out (Chapter 2), and may decrease the effectiveness of the explosion on the rock, whose physical-mechanical properties have become stabilized.

Thus, one factor in decreasing the effectiveness of mechanical drilling and the effectiveness of explosions at great depths is the equally worsening conditions for shattering rocks that are being acted on by hydrostatic (formational and rock) pressure.

Below we present data on the changes in physical-mechanical properties of rock and on the drilling characteristics as these depend on compacting loads; the work was done by E. I. Stetyukha [18].

Formational Pressure. The distribution of formational pressure at depth generally depends on various factors that are associated with peculiarities in the geologic structure of the rock. Table 9 shows the relationship between formational pressure and hydrostatic pressure at depths from 750 to 5000 m for a number of oil deposits displaying various geologic conditions.

From the cited data it may be seen that formational pressure generally exceeds hydrostatic pressure, especially at shallow depths (the exploratory drill holes in India).

At certain places the formation pressure at great depth may be less than hydrostatic pressure.

The Physical-Mechanical Properties of Rock at Various Depths. In the laboratories of the Grozneft (State All-Union Association of the Grozny Oil and Gas Industry) and the Krasnodarneft (Association of the Krasnodar Petroleum Industry) more than 10,000 cores of sedimentary rocks have been processed; these have been collected from Kuban, eastern Ciscaucasia, and the adjoining regions of Stravopol and the Crimea.

TABLE 9. The Relationship between Formational Pressure and Hydrostatic Pressure for Oil and Gas Deposits Exhibiting Various Geologic Conditions

Hole	Depth of hole, m	Ratio of formational pressure to hydrostatic pressure P_{form}/P_{hydr}
USSR		
Hole 2, Krainovka	2800	0.94
Hole 1, Ozek-Suat	3275	1.37
Hole 12, Karabulak	2000	2.07
Hole A, Klyuchevskaya	2000	1.05
Hole B, Klyuchevskaya	2065	1.00
Hole A, Kobystan	750	1.81
USA		
Formation:		
Temblor	1300	1.72
Bacon	2226	1.05
Hill	2257	1.03
Pittsburgh	2429	0.48
Bodcaw	2438	1.12
Horizon D-7	2800	2.07
Gulf region	3000	2.00
Waming well	3900	1.45
New Iberia	4362	1.42
2 AKS well	4573	1.40
Price No. 1 well	4657	1.42
20-1 ZKS well	4952	1.35
Austria		
Pitcher Creek	3600	0.91
India		
Exploratory No. 1	1200	3.00
Exploratory No. 2	1300	3.08

In studying these cores the following parameters were determined:

Average bulk weight γ, in g/cm^3;

Average density (mineralogical density), Δ, in g/cm^3;

Average porosity, B, in %;

Coefficient of recoil, ε, characterizing the elastic properties of the rock.

The parameters B, γ, and Δ are related in the following way:

$$B = 1 - \frac{\gamma}{\Delta};$$

$$\varepsilon = \sqrt{\frac{h_{rec}}{h}},$$

where h_{rec} is the height of recoil and h is the height of fall of the sphere to the surface of the sample.

The results of the core analyses are shown in Table 10; from these it follows that with increase in depth in bulk weight increases and the porosity decreases, but the average mineralogical density, as should be expected, changes very insignificantly. Thus for the Maikop rocks, with a change of depth approximately from 380 to 3900 m, γ changes from 1.84 to 2.47 g/cm^3, Δ from 2.74 to 2.77 g/cm^3, and B from 32.5 to 12.0%. In addition, the coefficient ε increases with increase in density and decrease in porosity.

TABLE 10. Changes in Physical Properties of Rocks with Depth

Age of rock	Depth of sample, m	Average bulk weight, γ, g/cm^3	Average mineralogical density, Δ, g/cm^3	Average porosity, B, %	Average coefficient of recoil, ε_{rec}
Cretaceous	1393—2370	2.37	—	7.2	0.59
	1944—2520	2.39	2.72	12.0	0.63
	2180—2600	2.46	2.74	10.0	0.66
Carboni-ferous	2370—2546	2.67	2.75	3.4	0.79
	2530—2536	2.72	2.73	3.0	0.72
	2660—2755	2.68	2.83	5.0	0.77
Elburgan (Paleocene)	1531—1944	2.26	2.71	16.0	0.57
	2037—2180	2.37	2.70	12.0	0.65
Sarmatian	195—253	1.90	2.73	30.2	0.37
	586—976	1.96	2.70	27.7	0.43
	1000—1240	2.02	2.74	26.0	0.41
Chokrak (Middle Miocene)	377—383	1.79	2.68	33.0	0.36
	987—1002	2.05	2.75	26.0	0.40
	1354—1371	2.13	2.78	26.2	0.44
Maikop (Middle Miocene)	383—761	1.84	2.74	32.5	0.39
	1070—1414	2.13	2.74	22.6	0.48
	1371—2074	2.26	2.78	17.9	0.70
Foraminiferal beds (Paleocene and Eocene)	761—805	1.96	2.75	29.6	0.38
	1414—1500	2.23	2.76	19.0	0.60
	2074—2479	2.36	2.68	11.7	0.59

Investigations, made by a number of workers, have shown that the strength of a single type of rock, characterized by some value of crushing strength (compressive strength, σ_{comp}), increases with increase in density. According to Borulya, El'yuninov, and Tsiskurelya, with a change in γ from 2.0 to 2.5 g/cm^3 the value of σ_{comp} for sandstones and limestones increases by a factor of 2.7-3.3. For igneous rocks σ_{comp} doubles with an increase in γ from 2.7 to 2.9 g/cm^3

The Specific Resistance of Rocks to Drilling. The specific resistance to drilling, according to V. S. Fedorov [48], is characterized by the coefficient σ, which generally increases with depth and is related not only to the physical-mechanical properties of the rock but also to the means of crushing them during drilling.

V. S. Fedorov has determined experimentally the values of σ for the sandy facies of the Novogroznenskii region, using a cutting bit during rotary drilling. The values of σ are expressed by the relations to the average value of the coefficient of recoil ε_{av}

$$\sigma = A(3.45\,\varepsilon_{av} - 1.32).$$

For easily drilled rock A = 1030 kg/cm^2; for rocks of moderate difficulty A = 1430 kg/cm^2, and rocks hard to drill have A = 2600 kg/cm^2.

M. L. Ozerskaya has proposed the following relationship for limestones:

$$\sigma = 500(\gamma - 1.58).$$

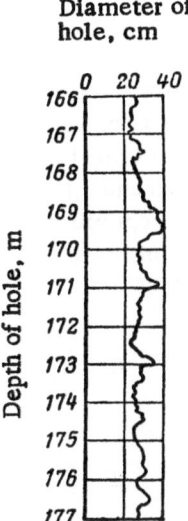

Fig. 36. Caliper log of a segment of the shaft of an experimental hole drilled by blasting in porous limestones and dolomites.

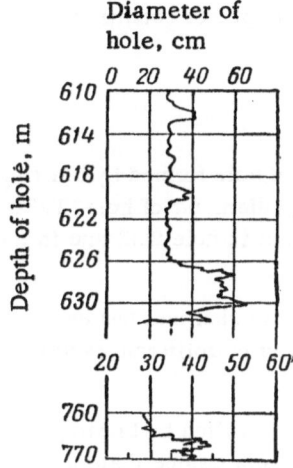

Fig. 37. Caliper log of segments of the shaft of hole 208 (Tuimazy region), drilled by blasting in very strong cherty limestones and dolomites.

For the clay rocks of eastern Ciscaucasia, σ increases approximately from 100 to 400 kg/cm² with increasing depth from 400 to 1600 m. The drilling rate of these rocks at great depths (3000-3500 m) drops by a factor of 7-20 as compared to the drilling rate at depths from 500-1000 m and in the surface layers.

From the considered data it follows that, for rocks of a single type, the bulk weight increases to some extent with increasing depth of occurrence, the porosity decreases, and the elastic properties (coefficient of recoil) increase. At greater depths there is also a marked increase in strength of the rock, which affects the rate of drilling and sometimes leads to a sharp retardation of the rate at great depths.

The Sinking of Deep Holes

From the results of experimental drilling in deep holes it is possible to trace the relationship between effectiveness of blasting and the ultimate diameter of the hole shaft in rocks of various susceptibilities to blasting.

Figure 36 shows the caliper log of a small (11 m) segment of an experimental hole (Moscow) drilled by detonating charges weighing 40 g in porous limestones and dolomites of Middle Carboniferous age; the bottom of the interval logged represents a depth of 177 m, and the average penetration per explosion was more than 20 mm.

The minimum diameter of the hole in the interval drilled by blasting is 230 mm, the average 310 mm. The maximum values, 370 and 400 mm, observed in zones through depth intervals of about 0.5 m, are not related to the explosive effect on the rock, since there are segments in the same hole, drilled by a bit 248 mm in diameter, where the walls are washed out to a diameter of 500 mm.

A similar explosive charge was tested in the depth interval 610-770 m (hole 208, Tuimazy region of the Bashkirian ASSR) during drilling in the Myachkovo and Podol'sk horizons of partly silicified dense limestones and dolomites with layers of chert, and also in Vereya limestones (locally argillaceous) with layers of clay.

A characteristic feature of the segments of hole 208 drilled by blasting (caliper log in Fig. 37, and Table 11) is the extensive widening (to more than 600 m), causing a sharp decrease in effectiveness of the blasting (down to 1.5 mm) as compared with the experimental hole (more than 20 mm) in much weaker limestones and dolomites.

The data obtained during the sinking of a hole by blasting in the Myachkovo and Podol'sk horizons, which consist of strong hard rocks uniform in character of shattering during blasting, show that a decrease in effectiveness of the explosion (from 4.7 to 1.5 mm) leads to an increase in average diameter of the hole (from 350 to 435 mm).

With the change in nature of the shattering by blasting and the decrease in explosive yielding of the rock of the Vereya horizon, caused by the transition to argillaceous limestones with layers of dense, highly plastic clays, the average diameter of the hole decreased to 340 mm, despite the lower effectiveness of the explosions (2 mm).

After increasing the directional effect of the explosions, improving the cleaning operation of crushed rock from the bottom of the hole, and diminishing the spreading of the charges over the bottom, the effectiveness of blasting was substantially increased, reaching (hole 855) an average of 12 mm in the Myachkovo and Podol'sk rocks, which are similar to the rocks in hole 208 and are uniform in character of shattering by explosions.

TABLE 11. Hole 208 (Tuĭmazy Region, Bashkirian ASSR)

Horizon	Rocks	Depth interval, m	Average penetration per explosion, mm	Average diameter of hole, mm	Rock group according to nature of shattering by explosion and relative susceptibility to blasting*
Myachkovo	Limestones with layers of dolomite, partly cherty		4.7	350	II
Podol'sk	Limestones, siliceous, with inclusions of chert	610-770	1.5**	435	II
Vereya	Limestones, locally argillaceous, with layers of dense clay		2.0	340	

* The classification of rocks by groups, as indicated in Tables 11-15, is discussed at the end of the chapter.

** The much lower effectiveness in Podol'sk rocks, as compared with Myachkovo, is explained by a change to higher frequency of explosions (from 340 to 680 per hr), leading to poorer conditions for cleaning the bottom of crushed rock fragments.

Figure 38 shows a short segment of a caliper log (560-568 m) typical of the section of hole formed by blasting in the depth interval 559-696 m. The effectiveness of the explosion and the average diameter of hole (Table 12) in the Myachkovo rocks are 13.3 mm and 306 mm (as against 4.7 mm and 350 mm in hole 208) and in the Podol'sk rocks 10.7 mm and 320 mm (as against 1.5 mm and 435 mm).

For testing the peculiarities of drilling holes by blasting in rocks that are much weaker and less dense, experiments were conducted in the same hole in the interval 1640-1650 m, in a bed of petroliferous sandstone, which was penetrated with an effectiveness of nearly 19 mm per explosion.

According to existing data, the diameter of the hole in sandstone in the segment drilled by blasting (Fig. 39) does not exceed 300 mm. Consequently, drilling by blasting in petroliferous sandstone that has a high susceptible to blasting does not lead to enlargements in the hole; rather it produces, with a higher penetration per explosion, a shaft with smaller diameter than those obtained in the considerably harder Myachkovo and Podol'sk rocks of the Tuĭmazy region.

Observations on the effectiveness of explosions and on the diameter of the hole produced have been made in clays and gypsum rocks of the Artinskian strata and in weaker limestones and dolomites of Upper Carboniferous, Myachkovo, Podol'sk, Kashira, and Bashkiria formations (an experimental-industrial hole, Kinel'-Cherkassy district of the Kuibyshev Oblast).

In comparing the effectiveness of charges in the hard rocks of the Myachkovo and Podol'sk strata (hole 855, Table 12) with the effectiveness of identical charges in the much weaker rocks of the experimental-industrial hole (Table 13), one may convince himself of the similarity of the average values, which agree satisfactorily with the results of laboratory investigations on models.

The explosive susceptibility of Upper Carboniferous porous, moderately tough limestones (see Table 13) proved to be greater (14.4 mm), but the smaller diameter of the hole (395 mm) should be explained by the erosive action of water, which is more intense than the erosive effect of clay mud.

TABLE 12. Hole 855 (Tuimazy Region, Bashkirian ASSR)

Horizon	Rocks	Depth interval, m	Average penetration per explosion, mm	Average diameter of hole, mm	Rock group according to nature of shattering by explosion and relative susceptibility of blasting
Myachkovo	Limestones and dolomites, cherty	559-625	13.3	306	II
Podol'sk	Cherty limestones	625-696	10.7	320	II
Naryshev-skii beds	Petroliferous sandstones and siltstones	1642-1651	18.8	300	I (chiefly) and III

TABLE 13. Experimental-Industrial Hole (Yablonovskii Area. Kinel'-Cherkassy District, Kuibyshev Oblast)

Horizon	Rocks	Depth interval, m	Average penetration per explosion, mm	Average diameter of hole, mm	Rock group according to nature of shattering by explosion and relative susceptibility to blasting
Artinskian	Clay, gypsum	729-747	6.7	345	III
Upper Carboniferous	Limestones of intermediate toughness, porous	756-1005	14.4	395	II
Myachkovo	Dolomites	1009-1207	9.9	355	II
Podol'sk	Dolomites and limestones	1210-1435	8.5	360	II
Kashira	Limestones and dolomites	1467-1492	8.9	350	II
Bashkiria	Limestones	1515-1538	9.5	330	II

Explosive drilling in plastic clays and gypsum rocks of the Artinskian horizon, which represent another group according to the nature of shattering by explosion and which possess minimum yielding to explosion, was accompanied by a great decrease in effectiveness without any substantial diminution of diameter, as compared with explosive drilling in hard, brittle rocks.

Explosive drilling was carried out in siliceous limestones and sandstones with layers of clay (of the Tournaisian series), in siliceous limestones with layers of clay (of the Famennian series), in dolomitized siliceous limestones with layers of clay and in marls (of the Frasnian series), and in clays (of the Shugurovo formation).

From the data in Tables 14 and 15 it may be seen that a decrease in effectiveness of explosions in rocks of the Famennian series, to 4.5 mm, led to enlargements in the hole, with a maximum diameter of 700 mm, and to an increase in the average diameter of the hole, to 500 mm. However, with some increase in effectiveness of the explosions (6 mm) the average diameter of the hole decreased (to 350 mm).

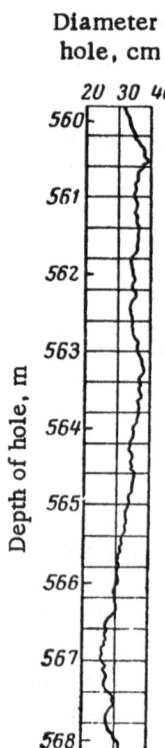

Fig. 38. Caliper log of segment of hole 855 (Tuimazy region), drilled by blasting in very strong cherty limestones and dolomites.

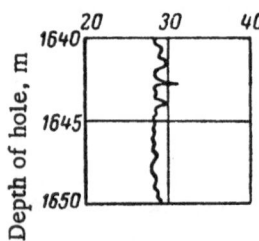

Fig. 39. Caliper log of segment of hole 855 (Tuimazy region), drilled blasting in petroliferous sandstone.

In evaluating the explosive susceptibility of the rocks of the Famennian series it may be assumed that the numerous alternations of clay layers impairs the drilling characteristics of cherty limestones as compared with conditions in a continuous sequence of such limestones (such as the cherty limestones of the Podol'sk formation in the Tuimazy district). One should also keep in mind that plastic clays and elastic-brittle limestones have greater strength and resistance to explosions at great depths because of compaction by hydrostatic and rock pressure.

In addition, the shattering of rocks by explosions (especially hard rocks) on the bottom and on the walls of a hole takes place under different conditions. The shattering of rock on the bottom of a hole is due chiefly to the mechanical action of compressional and dilational waves propagated through the mass

TABLE 14. Hole 1D (Yablonovskii Area, Kinel'-Cherkassy District, Kuibyshev Oblast)

Horizon	Rocks	Depth interval, m	Average penetration per explosion, mm	Average diameter of hole, mm	Rock group according to nature of shattering by explosion and relative susceptibility to blasting
Tournaisian	Cherty limestones and sandstones with layers of dense clay	2235-2287	4.5	Not measured	II (chiefly) and III
Famennian	Cherty limestones with layers of dense clay	2320-2349	6.0	350	II (chiefly) and III

by direct transmission of energy from the compressed products of the explosion. The shattering of rocks on the walls of the hole takes place chiefly by repeated development of positive and negative stresses, arising through explosive disturbances propagated through the liquid during the explosions.

The extension of fractures and the development of spalling in hard rocks are more pronounced on the walls of a hole than on the bottom.

The lowest effectiveness of explosions is found in the compacted clays and marls of the Frasnian series and of the Shugurovo formation.

TABLE 15. Hole 2D (Yablonovskii area, Kinel'-Cherkassy Petroleum District, Kuibyshev Oblast)

Horizon	Rocks	Depth interval, m	Average penetration per explosion, mm	Average diameter of hole, mm	Rock group according to nature of shattering by explosion and relative susceptibility to blasting
Famennian	Cherty limestones with layers of dense clay	2406-2629	4.5	500	II (chiefly) and III
Frasnian	Dolomitized limestones with layers of dense clay, marl	2630-2711	3.2	Not measured	II and III
Shugurovo formation		2799-2807	3.7	Not measured	III

The comparatively low effectiveness of explosions and the extensive enlargements of the hole at great depth cannot be explained merely by the compaction of the rock and the diminution of explosive susceptibility. Hydrostatic (and rock) pressure may also be of fundamental significance; this affects not only the strength characteristics of many rocks but also the mechanism of shattering of these rocks by explosion.

The general tendency of decrease in penetration per explosion with depth points to an increase in the diameter of the hole, the change of which, in relation to depth, is shown graphically in Fig. 40.

On the basis of investigations on the formation of holes by blasting, on models and on actual outcrops, a classification of rocks is proposed in accordance with the nature of their shattering by explosions and with their relative susceptibility to explosions [17].

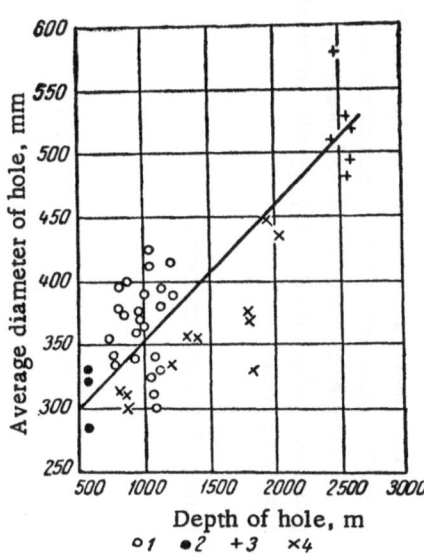

Fig. 40. Change in average diameter of hole in relation to depth (according to data on explosive drilling). 1) Experimental-industrial hole (Yablonovskii area); 2) hole 855 (Tuimazy district); 3) hole 2D (Yablonovskii area); 4) hole 277 (Mukhanovo area).

Rocks may be divided into three typical groups according to the nature of their shattering by the direct effect of explosions and by explosive disturbances transmitted through liquid:

The first group consists of soft and weak, uncompacted rocks, whose shattering by explosions is characterized chiefly by crumbling;

the second group consists of brittle, hard, and dense rocks that, when acted on by explosions, are mainly crushed in the shatter zone of the explosion and spall off in comparatively large pieces in the hydraulic zone;

the third group, according to nature of shattering, consists of highly plastic rocks. These rocks are less susceptible to the development of fractures and spalling than the second group. The effects are manifested in much smaller fragments separated from the mass on the bottom of the hole (direct effect of the compressed products of the explosion) and from the walls of the hole (derived shock waves transmitted by the liquid).

The explosive shattering of rocks of the third group is characterized by slight crushing on the bottom of the hole and by weakly defined spalling of comparatively small fragments from the walls of the hole.

TABLE 16. Approximate Classification of Rocks according to the Nature of Shattering by Explosion and to Relative Susceptibility to Blasting

Rock group according to nature of shattering by explosion and relative susceptibility to blasting	Brief description of rocks	Nature of rock shattering by blasting	Dispersal of rock fragments broken off by blasting*	Relative susceptibility of rocks to blasting
I	Friable, soft, and weak uncompacted sedimentary and metamorphic rocks occurring relatively near the surface, and also permeable rock not subjected to the compaction of hydrostatic and rock pressure (such as sandstone) occurring at various depths.	Chiefly crumbling	Greatest	Greatest
II	Brittle, hard, dense sedimentary rocks (such as limestones and dolomites) occurring chiefly at great depths, and very dense, brittle, tough igneous rocks (such as granite and diabase) occurring at various depths.	Crushing and spalling	Coarse; average weight of fragments greater than 20–30 g	Moderate
III	Sedimentary rocks of low and intermediate toughness with closed pore channels, compacted at great depths by hydrostatic and rock pressure, possessing high plasticity (clay, gypsum, siltstone, anhydrite, etc.)	Crushing and poorly defined spalling	Fine; average weight of fragments less than 10 g	Least

*Obtained by detonating charges of 50 g of explosives.

The explosive yielding of rocks belonging to the first group in reference to shattering characteristics (crumbling) is the greatest of the three groups.

A wide range of properties in rocks of the second and third groups in reference to shattering characteristics (crushing and spalling) leads to substantial variations in explosive yielding within each group and even in one particular kind of rock under different conditions, depending on compaction of the rock by hydrostatic, formational, and rock pressure.

An approximate classification of rocks according to the kind of deformation by explosion and to the relative susceptibility to blasting is shown in Table 16 and in Tables 11-15, illustrating the mutual relationship between effectiveness of explosions and the diameter of a hole.

Chapter 4

THE WORK OF EXPLOSIVE CHARGES
DURING UNDERWATER EXPLOSIONS

Solid or liquid explosives may be used for explosive drilling. Liquid explosives have several advantages over solid.

1. Liquid charges are prepared automatically during explosive drilling from chemical components that are not explosive in their initial state. The transportation, preservation, and use of these components are not complicated by any special requirements or standard specifications that apply to ordinary unitary explosives and to the means of initiating them. Thus an operation free from danger and accident is secured.

2. The adopted liquid explosives are nonhygroscopic, are insoluble in water, and are safely detonated at high initial pressures, reaching 500-1000 kg/cm^2 in deep holes.

3. The ignition of the liquid explosive charge (under normal conditions), because of its high sensitivity, may be effected by a thermal impulse from an ignition capsule that is not dangerous to handle, whereas the ignition of a standard solid explosive charge is produced by a detonator that is dangerous to handle and is also more expensive.

4. The power of certain liquid explosives and their shattering effect are greater than for standard solid explosives.

5. The use of liquid explosives has facilitated the pursuit of numerous investigations, at the stage of experimental operations, with charges of various sizes and shapes.

The explosive process of drilling holes is carried out with charges of powerful, chemically stable liquid explosive. The principal properties of some liquid explosive mixtures are shown in Table 17.

For a hundred thousand explosions it was established that the effectiveness (penetration per explosion) decreases when a hole reaches the depth of 1200-1500 m. The decrease in effectiveness of blasting with increased depth of drilling may be explained by the influence of certain constantly effective factors, of which hydrostatic, formational, and rock pressures are directly related to depth of the hole. Below we consider the principal results of experimental studies on preliminary evaluations of the effect of hydrostatic pressure and other factors on the operation of explosive charges.

Hydrostatic Pressure and the Operation of the Explosive Charge

An attempt is made to distinguish fougasse and brisance forms of mechanical work performed by an explosion. The fougasse form is characterized by the action of the expanding gaseous products of the explosion and of the shock wave that is propagated through the medium surrounding the charge; such action is effective at some distance from the surface of the medium and is defined by the culminating work of expansion of the explosion products. The magnitude of this effect depends chiefly on the potential energy of the explosive charge, on the specific volume and thermal capacity of the gaseous products of the explosion, and, as computations show, it changes markedly with any change in ultimate pressure, up to which value the explosive products may continue

TABLE 17. Properties of Some Liquid Explosive Mixtures

Principal properties	Mixture 1	Mixture 2	Mixture 3	Mixture 4	Mixture 5
Density, g/ml	1.49	1.44	1.48	1.49	1.51
Heat of explosion					
kcal/kg	1 720	1 480	1 080	1 390	1 170
kcal/liter	2 500	2 160	1 600	2 070	1 770
Specific volume of explosive products, liter/kg	770	710	640	705	715
Velocity of detonation, m/sc	7 800	7 000	6 700	7 200	6 430
Velocity of detonation products, m/sec	1 950	1 750	1 670	1 800	1 605
Pressure at front of detonation wave, kg/cm^2	227 000	167 000	168 000	196 000	156 000

TABLE 18. Brisance Effect of Explosions at Various Initial Pressures in the Medium Surrounding the Explosive Charges

Initial pressure, kg/cm^2	Reduction of lead cylinder				Average brisance	
	min., mm	max., mm	average			
			Δh. mm	%	$\dfrac{\Delta h}{h_0{}^* - \Delta h}$	%
1	11.1	12.5	11.8	100	0.876	100
100	11.3	11.8	11.6	98.5	0.851	97.2
150	11.2	12.2	11.6	98.5	0.856	97.7
200	10.8	11.6	11.3	96.0	0.816	93.2
300	10.2	12.4	11.1	94.4	0.792	90.5
400	10.6	11.4	11.0	93.5	0.772	88.3

* h_0 represents the initial height of the lead column.

to expand. Thus, if we take 100 to represent the work of expansion of the explosive products up to atmospheric pressure, the work of expansion will be 80 if the ultimate pressure in the medium surrounding the charge is 100 kg/cm^2, and 76 if this value is 200 kg/cm^2.

The brisance form of the explosion is characterized by the direct effect on the solid mass of the compressed products of the explosion at high pressures and density; the energy of this phase is but a comparatively small part of the total work of the explosion.

The brisance effect reaches a maximum when the explosive charge is in direct contact with the rock mass in the direction of propagation of the detonation wave. The density of the explosive charge has a substantial effect on the brisance properties, since the magnitude of the detonation pressure depends to a great degree on the initial density of the charge. At increasing distances from the source of the explosion the mechanical effect of the explosion decreases sharply because of the abrupt diminution in pressure, velocity, and density of the explosion products.

Theory indicates that when there is spherical symmetry the pressure in the explosion products decreases in inverse proportion to r^9, where r is the distance from the source of the explosion.

Both forms of explosive work play substantial roles in the formation of a hole.

The fougasse effect appears in the zone of hydraulic effect of the explosion, brisance in the zone of crushing. The fougasse effect chiefly determines the future expansion of the zone of crushing of succeeding explosions, up to the formation of a hole of established diameter.

The brisance, or local, effect of an explosion depends on the maximum pressure at the front of the detonation wave, P_m, and also on the velocity of the detonation, D; these are related by the well-known equation

$$P_m = \frac{1}{4} \varrho_0 D^2, \tag{9}$$

where ρ_0 is the density of the explosive charge.

From relation (9) it may be seen that a change in the velocity of the detonation may markedly affect the maximum pressure and, consequently, the local (or brisance) action of the explosion.

According to existing views concerning the mechanism of ignition and concerning the course of detonation of an explosive charge, the increased pressure in the medium surrounding the charge, under certain circumstances associated with the physical-mechanical properties of the explosive charge and of the initiator of the explosion, may decrease the susceptibility of the charges to detonation. It is therefore necessary to investigate the effect of hydrostatic pressure on some of the properties of explosive charges.

During laboratory determinations of the effect of initial pressure on the duration of an explosion of liquid explosive, two types of experiments were performed. Some of the experiments were conducted with a delay detonator connected to the model charge in a thick-walled container in which the required pressure was supplied. The container was impelled along special guides against the rigid face of the rock mass; during this operation an oscillograph recorded the time between the blow of the container against the rock and the instant of explosion. These experiments showed that the duration of the explosion at pressures of 1 and 100 kg/cm² within the container is the same for each, and amounts to about $1 \cdot 10^{-3}$ sec.

In other experiments the ignition of the explosion and the completeness and velocity of the detonation were evaluated by the brisance effect of the explosive at various initial pressures in the medium surrounding the charge.

In these experiments a method was employed that is basically similar to the standard method, but in which the dimensions of the liquid explosive charges and of the lead column were decreased in proportion to the strength of the container. The results of the experiments (Table 18) have shown that the relative deformation of the lead column, serving as a measure of the "true brisance" (according to M. A. Sadovskii), decreased only 12% on changing the initial pressure from 1 to 400 kg/cm².

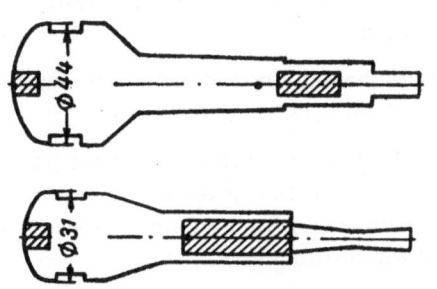

Fig. 41. Diagrams of charges of liquid explosives of 33 ml (50 g) and 60 ml (90 g).

Further experiments were conducted in deep holes with two actual charges of liquid explosives having similar shapes, but differing in size. Figure 41 shows diagrammatically a charge of liquid explosive 33 ml in volume (weight of 50 g) having the form of a body of rotation and a ratio of height (to point of initiation) to diameter of charge of 1.88 : 1; the hemispherical head contained about 60% of the mass of the explosive and the initiator was placed in the upper part. The figure also shows a charge approximately similar in form to the first, with h : d = 2 : 1, having a greater weight (90 g), and also with the initiator in the upper part, 80% of the explosive mass being concentrated in the hemispherical head (the elements of design of the capsules are shown by crosshatching within the charges).

Explosive mixture 1 was used in the experiments with the indicated charges (see Table 17).

The effect of hydrostatic pressure on the brisance properties of explosives is determined by the deformation of steel disks (175 mm in diameter and 8, 10, and 12 mm thick), which are immersed in the hole at various depths, let down on pipes in a special device. The bending of a disk immediately under the charge, the piercing of the disk, and other phenomena of the brisance effect of explosives apparently depend on the impulse communicated to the disk by the explosion products, which have a very high pressure (more than 200,000 kg/cm² for the liquid explosive used).

In analyzing the results of the tests the depth of indentation and the diameter of perforation were taken into account.

Table 19 shows the averaged results of experiments with steel disks, the deformation of which may serve to indicate changes in the brisance properties of an explosive (and the local effect of an explosion) as affected by hydrostatic pressure in a hole with an explosive charge (50 g) and disk immersed to a depth of 175-1450 m in water.

In the indicated experiments the effect of an explosion as evaluated by the indentation and perforation of steel disks changed little with hydrostatic pressure. In several experiments the deviation was no greater than found in parallel experiments at a single depth of immersion of the disks.

TABLE 19. Local Effect of an Explosive Charge (50 g) at Various Hydrostatic Pressures in the Medium (Water) Surrounding the Charge *

Hydrostatic pressure, kg/cm^2	Deformation of steel disk	
	indentation, mm	diameter of perforation, mm
17.5	31	22
70.0	29	20
145.0	28	20

* Material in envelope of charge was plastic (fiber).

Similar investigations made in a hole filled with drilling mud (density of 1.2 g/cm^3) gave results contradictory to the experiments in the hole filled with water. In these tests the greatest deformation of the steel disks (open perforation) was found in disks immersed at the shallowest depths in the drilling mud (immersion depth of 650 m, hydrostatic pressure of 80 kg/cm^2), and the least deformation (indentation of 24 to 27 mm) was found in disks most deeply immersed (immersion depth of 2400 m and a pressure of 300 kg/cm^2). In this drilling mud the disks ceased being pierced at immersion depths greater than 800 m (about 100 kg/cm^2 pressure), whereas disks immersed in water to a depth of 1450 m still became perforated.

A check on the parameters of the drilling mud circulating in the hole showed a considerable fluctuation in the viscosity (from a value of 20 sec to a nonfluid state) and in density (from 1.12 to 1.26 g/cm^3).

Because the consistency of drilling mud occurring beneath a relatively thin steel disk is inconstant (ranging from water to packed mud), owing to conditions in the hole, the local effect of an explosion also proves to be somewhat modified by the density of the medium supporting the disk (membrane) from the rear; experiments were conducted with massive lead targets (120 mm high) in a hole filled with drilling mud.

Figure 42 illustrates a typical deformation of lead targets exposed to an underwater explosion of similar charges of liquid explosive (about 50 g). The brisance cavity is a somewhat "eroded" cone, in which, however, the diameter of the entry aperture and the depth may be measured with sufficient precision.

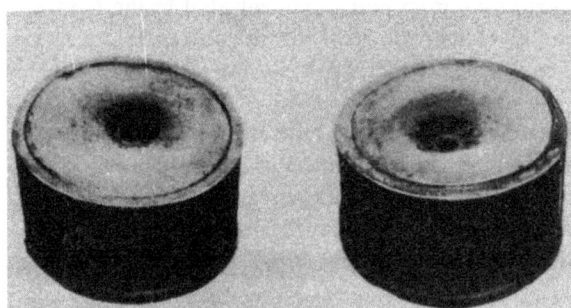

Fig. 42. Typical deformation of lead targets during underwater explosion of superincumbent explosive charges of 33 ml (50 g).

Table 20 shows typical results of a series of experiments with lead targets in holes filled with drilling mud and with water. These experiments have confirmed the invariability of the mechanical effect of an explosion with changes in hydrostatic pressure acting on the explosive charge, for pressures ranging from 75 to 300 kg/cm^2. The results are in good agreement with the experimental data for steel disks (see Table 19), and they also point out the fact that the properties of an explosive are practically invariant during changes in hydrostatic pressure up to about 300 kg/cm^2.

In the described experiments the charge of liquid explosive and the initiator were placed in a plastic envelope,

TABLE 20. Local Effect of an Explosive Charge (50 g) at Various Hydrostatic Pressures in the Medium Surrounding the Charge*

Target No.	Surrounding medium	Immersion depth, m	Hydrostatic pressure, kg/cm²	Deformation of lead target		
				depth of cavity, mm	entry diameter of cavity, mm	volume of cavity, cm³
54	Drilling mud, density of 1.2-1.22 g/cm³, viscosity of 25-30 sec	2400	300	19	44	25
58	The same	2400	300	19	50	26
60	" "	2400	300	21	50	29
56	" "	650	80	19	45	26
57	" "	650	80	19	44	29
59	" "	650	80	19	46	26
62	Water	750	75	18	44	25
65	The same	750	75	18	45	25

* Material in envelope of charge was plastic (fiber).

the inner cavity of which communicated with the surrounding liquid medium, permitting the hydrostatic pressure in the hole to be transmitted to the charge and the initiator.

For a more complete study of the effect of hydrostatic pressure on the work of an explosive charge, comparative experiments were conducted with hermetically sealed charges and initiators. For this purpose a constant explosive charge and initiator (capsule igniter) were placed in a strong steel envelope, isolating them from the pressure of the surrounding medium. This arrangement created general conditions for exploding the charge at atmospheric pressure.

In comparative experiments for communicating the pressure of the surrounding medium to the charge in the metallic container, the seal was removed or a small hole was drilled into the container.

Typical examples of deformation in lead targets for hermetically sealed charges and for charges communicating with the surrounding medium are indicated in Table 21.

The experiments with lead targets have established that hermetic sealing of the charge of liquid explosive and of the initiator within a strong steel container, isolating them from any effect of the hydrostatic pressure of the surrounding medium, does not enhance the local effect of an explosion, as compared with a charge placed in a similar container, but with the inside communicating with the surrounding medium. Consequently, no effect of hydrostatic pressure (up to 300 kg/cm²) on the brisance properties of an explosive was detected when hermetically sealed charges and initiators were used.

Similar results were obtained in experiments on steel disks with identical charges placed in steel containers.

Experiments with charges of liquid explosive in containers of strong aluminum alloy, of various plastics, and of a mixture of iron powder and plastic have shown that these materials have but an insignificant effect on the form and size of cavity produced in the lead target. A substantial increase in size of cavity observed in experiments with a charge in a steel container is explained by the strong steel cover (forming the head of the hemispherical charge), the mass of which is set in motion by the explosion products, causing an increase in the local effect (especially as compared to lead).

In order to avoid misrepresentations that may possibly arise because of the plasticity of lead, control experiments were set up with massive Duralumin targets (160 mm in diameter, 65 mm high). These experiments showed that with a greater than fourfold change in hydrostatic pressure (from 70 to 300 kg/cm²) the local effect of the explosion, as in the experiments with lead targets, remained practically unchanged. This also points to stability of the properties of the explosive within the investigated range of pressures.

TABLE 21. Local Effect of Explosive Charges (50 g) with the Charge Hermetically Sealed and with the Charge Communicating with the Surrounding Liquid Medium, Which Is under Hydrostatic Pressure

Surrounding medium	Immersion depth, m	Hydrostatic pressure, kg/cm²	Pressure communicated to explosive charge, kg/cm²	Deformation of lead target			
				form of cavity	depth of cavity, mm	entry diameter of cavity, mm	volume of cavity, cm³
Drilling mud, density of 1.2-1.22 g/cm², viscosity of 25-30 sec	2400	300	Atmospheric	Hemisphere	28	65	57
The same	2400	300	Hydrostatic	The same	27	70	59
" "	2400	300	The same	Ellipsoid	31	63	43
" "	800	100	Atmospheric	Hemisphere	25	65	50
" "	650	80	The same	The same	28	63	50
Water	750	75	Hydrostatic	" "	24	63	50

TABLE 22. Local Effect of an Explosion of Greater Explosive Charge (90 g)*

Target	Surrounding medium	Immersion depth, m	Hydrostatic pressure, kg/cm²	Deformation of target (average of 10 expts.)			
				nature of deformation	indentation in steel disk and depth of cavity in lead, mm	entry diameter, mm	vol. of cavity, cm³
Steel disk	Water	800	80	Perforation	35	42	
Lead target	Drilling mud, density of 1.2 g per cm³, viscosity of 30 sec	650	80	Conical cavity	24	60	46

* Material in envelope of charge was made of plastic (fiber).

The average deformation (of a series of experiments) of the steel disks and lead targets by explosions when the charges, of the same explosive, were increased to 90 g are shown in Table 22.

In the lead targets the shapes of the cavities obtained by exploding charges of 90 and 50 g were similar. The ratio of the depths of the cavities is approximately proportional to the cube root of the ratio of weights of charges, and the volume of the cavity is approximately proportional to the weights of the charges.*

In experiments with a charge of 20 g, having a form similar to the charges of 45 and 90 g, results were obtained that exhibit the same relationship.

* The data are compared in Tables 22 and 20.

The parameters of a charge, the intensity of the initiator, and the position of the initiator within the charge effect to a considerable degree the local effect of an explosion, and on this the depth and aperture of the funnel of crushing depend.

Experiments on models (see Chapter 2) have shown that for a cylindrical charge 4r in height (r being the radius of the charge) a shift of the "point" of initiation from the base to the top may increase the penetration per explosion in hard rocks by a factor of approximately 2.5. This has been confirmed experimentally during ignition of detonations in highly sensitive liquid explosive by a relatively weak impulse of the "point" initiator of an explosion. By a "point" initiator we mean one with very low energy and a small zone through which the initiating impulse is propagated, as compared with the energy and magnitude of the rupturing charge.

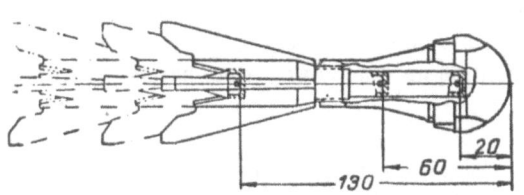

Fig. 43. Diagram of an experimental capsule with various positions of the "point" initiator of the explosion according to height of charge.

Figure 43 illustrates an experimental capsule and indicates three different positions of the initiator (capsule igniter KV-1) according to height of charge (20, 60, and 130 mm).

Tests made under a layer of water have shown that a shift of the point initiator from the lower part of a change (height of initiation, 20 mm) to the upper part (height of initiation, 130 mm) also leads to a substantial increase in the local effect of the explosion (see Table 23).

During the detonation of a liquid explosive charge by a comparatively powerful KD-8 capsule detonator, it was found that a higher position of the detonator in the charge, beginning with a height equal to the radius of the charge, had no effect on the local action of the explosion. This may be explained by the zone of initiating influence of the capsule detonator being commensurate with the dimensions of the charge of highly sensitive liquid explosive.

With further increase in the power of the detonator the local effect of the explosion is maintained at a level represented by a similar charge having the disadvantage of a low position (20 mm) of the point initiator (KV-1).

The height of the cylindrical charge with the upper position of the initiator was varied from 4r to 8r, and in addition, the power of the capsule igniter was doubled.

Experiments were conducted at hydrostatic pressures up to 300 kg/cm^2, and they demonstrated that increasing the length of the cylindrical charge beyond 4r with initiators of various strengths placed in the upper position leads to no further increase in the local effect of the explosion.

The results of these experiments are found to be in agreement with the theory concerning the active part of a charge as worked out by Vlasov and Pokrovskii and further developed in the works of Baum and Stanyukovich [49, 50].

The active part of the charge is considered to be the detonation products emitted in a given direction.

According to this theory, which has been confirmed by experimental investigations, the active mass of the charge increases with increase in length of the charge (for a given diameter) only up to a certain limit. This limiting value for the active mass of a charge of given diameter is reached when the length $l_{lim} = \frac{9}{2} r$. From this it follows that when the length of a charge is increased beyond the optimum value, $\frac{9}{2} r$, the specific impulse of the explosion, which defines the brisance effect in a first approximation, does not increase. The specific impulse I_0 depends on the velocity of the detonation, the density of the explosive, and the weight and geometry of the charge:

$$I_0 = \frac{8}{27} \varrho_0 l_{lim} D = \frac{8}{27} MD,$$

where ρ_0 is the density of the explosive, l_{lim} is the length of the active part of the charge, and M is the mass of the charge.

Consequently, the impulse increases linearly with increase in velocity of the detonation, and it may be increased by increasing the density of the explosive.

Equation (10) also shows a linear relationship between the impulse and the length of the active part of the charge, but this relationship is not actually observed, since it is practically impossible to achieve one-dimensional movement of the detonation products, to prevent completely any lateral movement even when the charge is enclosed in a strong container. The relationship in (10) may be used for three-dimensional movement of the detonation products if the entire mass of the charge is changed into the mass of the active part, the value of which is computed for each individual case.

Table 23 shows the average values of deformation of steel disks, values obtained in determining the local effect of an explosion under a hydrostatic pressure of 300 kg/cm^2 with cylindrical charges of liquid explosive of approximately equivalent weights, but with initiators of various strengths placed at various heights within the charge.

TABLE 23. Local Effect of Explosions of Explosive Charges (50 g) Depending on Position and Strength of Initiator under a Hydrostatic Pressure of about 300 kg/cm^2 in the Surrounding Medium (drilling mud)

Height of initiator position in charge, mm	Primer	Thickness of steel disk, mm	Deformation of steel disk (average value of a series of experiments)		
			nature of deformation	diameter of perforation, mm	depth of indentation, mm
20	KV-1	8	Indentation	—	23
60	The same	8	Perforation	22	—
20	" "	12*	Indentation	—	19
60	" "	12	The same	—	27
60	KV-2	12	" "	—	28
130	The same	12	" "	—	27
20	KD-8	8	" "	—	22
60	The same	8	" "	—	23
60	Detonator 1 g	8	" "	—	22
60	The same 4 g	8	" "	—	24

* During experiments with steel disks 12 mm thick, observations were increased (as compared with disks 8 mm thick), since the deformation of such disks by explosion was not accompanied by open perforation.

The data in the table confirm the following:

1) The greatest (and practically identical) local effect of an explosion is achieved when the point initiator is placed on the upper part of the charge (60 and 130 mm); it does not depend on the strength of the capsule igniter (KV-1, KV-2).

2) The local effect of an explosion during ignition by a detonator does not depend on the strength of the detonator or its position in the charge; it corresponds to the minimum result obtained when a "point" initiator (KV-1, KV-2) is placed in a low position (20 mm) in the charge.

In order to evaluate the influence of distribution of the explosive mass on the local effect of an explosion, comparative experiments with three charges of equivalent weight (75 g) of highly sensitive liquid explosive were performed; the point initiator was placed on top, and the explosions were set off under a layer of water on steel disks 18 mm thick.

Figure 44 shows schematically three charges of equivalent weight with different concentrations of explosive mass in the head. In the first and second examples (a and b) the initiator was placed at distances of 93 and 86 mm from the bottom, respectively; in the third charge (c), the distance was 55 mm.

Figures 45a, b, and c, respectively, show the characteristic deformation in the steel disks caused by explosions of the indicated charges. The greatest deformation of a disk (Fig. 45b) was produced by an elongated charge, the principal mass of which (80%) was concentrated in the head (Fig. 44b); the least deformation (Fig. 45c) resulted from a short concentrated charge (Fig. 44c). The deformation due to the charge illustrated in Fig. 44a was intermediate (Fig. 45a).

The influence of the distribution of explosive mass (according to height of charge) on the local effect of an explosion may also be explained by the theory of the active part of the charge. The active part of a charge increases with the specific impulse of the explosion when the diameter of the charge is increased, asymptotically approaching a definite limit.

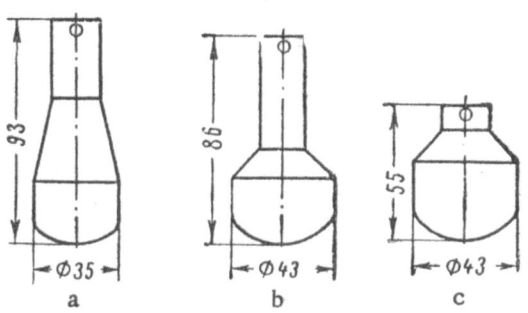

Fig. 44. Diagrams of charges of equivalent weight (75 g) with different concentrations of explosive in the head.

Influence of the Medium Surrounding the Charge and the Solid Face on the Action of the Explosion

The influence of the medium surrounding the charge and the face of the mass on the brisance and fougasse effects of an explosion was tested on relatively thin steel disks and on massive lead targets by explosions in air, in water, and in clay and cementing muds of various consistencies.

At a constant impulse of explosion the magnitude and nature of deformation in a steel disk and in a lead target

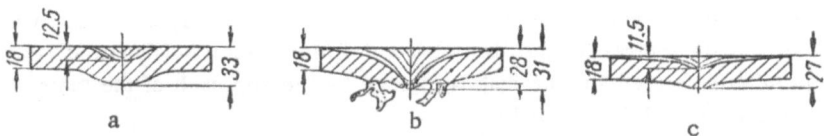

Fig. 45. Local effect of underwater explosions on steel disks by charges (75 g) variously concentrated in the head.

depend on the density and compressibility of the medium surrounding the disk (on all sides), the lead target (only on the side of the charge, because of its massiveness), and the explosive charge.

The influence of the density and viscosity of the medium surrounding the charge and the steel disk on the local effect of the explosion was tested on a stationary base with charges of liquid explosive weighing 50 g (mixture 2).

Good reproducibility was obtained in the experiments on steel disks. The results of these experiments are shown in Table 24, the data for two disks (those showing maximum and minimum deformation) being given for each type of medium.

It may be seen from Table 24 that explosions in air produce perforations of the steel disks. If the disk is surrounded on all sides by water, the indentation of the disk increases somewhat without actual perforation occurring. The over-all deformation of the disk decreases with change of the surrounding medium from water to drilling mud.

A further considerable increase in density (up to 1.8 g/cm^3) and in viscosity (above 60 sec) of the medium leads to no marked decrease in deformation of the steel disks, as compared with deformation in drilling mud.

Further, in investigating the influence of the surrounding medium on the local effect of an explosion (Table 25), the following experiments were performed:

1) A layer of water was placed over the steel disk and the explosive charge, and air was placed against the back of the disk; the characteristic deformation of the disk is shown in Fig. 46a.

TABLE 24. Local Effect of an Explosion of a Charge (50 g) in Relation to Properties of the Surrounding Medium

Surrounding medium	Deformation of a steel disk*	
	indentation, mm	diameter of perforation, mm
Air	21	12
	24	12
Water	27	—
	31	—
Drilling mud, density of 1.23 g/cm³, viscosity of 35 sec	23	—
	24	—
Cementing mud, density of 1.8 g/cm³, viscosity greater than 60 sec	22	—
	24	—

*When there was damage to the base of the explosive charges on steel disks 8 mm thick, immersed in liquid, the disks were not pierced, in contrast to the "dynamic" tests (see Table 19). This is explained by the fact that during damage to the base of the head of the charge, having the form of a hemispherical lid (of plastic), is in contact with the disk at only one point, whereas, when the lid approaches the disk with a definite velocity ("dynamic" tests), the lid is deformed and the disk is pierced by the approach of the mass of the charge to the solid face (disk).

2) The surrounding medium was water exclusively (Fig. 46b).

3) Air was placed above the disk and charge, with water at the back of the disk (Fig. 46c).

4) For comparison the disk and the charge were placed in air (Fig. 46d).

A comparison of the experimental results shows a very substantial increase in the over-all deformation of a steel disk when there is air at the back side of the disk (Fig. 46a), as compared with the local effect when the

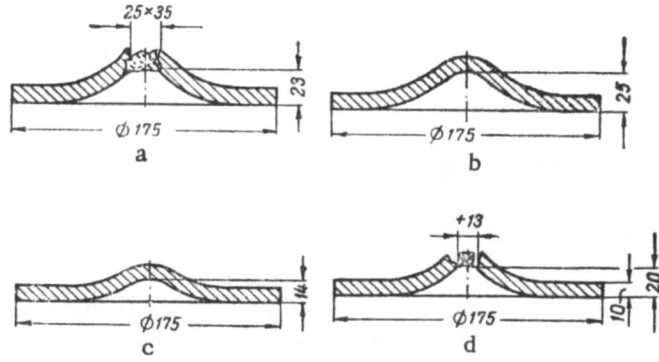

Fig. 46. Local effect of an explosion on a steel disk with various arrangements of liquid and air relative to the disk and the explosive charge (50 g). a) A layer of water above the disk b) water all around; c) air above the disk and charge, water behind; d) air all around.

TABLE 25. Local Effect of Exploding a Charge (50 g) for Various Arrangements of Liquid and Air Relative to the Disk and the Charge

Surrounding medium	Deformation of steel disk* (average of five expts.)		
	nature of deformation	diameter of perforation, mm	depth of indentation, mm
Water above the steel disk and the charge, air at back of disk	Perforation (Fig. 46a)	25 × 35	23
Water	Indentation (Fig. 46b)	–	25
Air above disk and charge, water at back of disk	Indentation (Fig. 46c)	–	14
Air	Perforation (Fig. 46d)	13	20

*Disk 10 m thick, 175 mm in diameter.

disk is completely immersed in liquid (Fig. 46b). The least deformation of the disk is noted for the experiments in which the charge was surrounded by air and the back side of the disk was against water (Fig. 46c).

Thus, in a medium denser than air, hindering the expansion of the gaseous explosion products at the initial moment and being partly drawn into motion by these products, the active part of an explosive charge is somewhat increased, and, consequently, so is the unidirectional impulse of the explosion and its local effect.

From the experiments with steel disks it may be concluded that the medium at the back of a disk, depending on its density and viscosity, proves to have a variable retarding effect on the flow of the detonation products. This is confirmed, for example, by a sharp drop in the spalling effect when an explosion is discharged under water instead of in air.

The influence of density and viscosity of the medium on the local effect of an explosion was also tested with massive lead targets (Table 26).

TABLE 26. Local Effect of Exploding Charge (50 g) with a Surrounding Medium of Various Densities and Viscosities

Surrounding medium	Deformation of lead target (average of a group of expts.)		
	depth of cavity, mm	entry diameter of cavity, mm	volume of cavity, cm^3
Air	17	42	25
Water	18	45	29
Drilling mud, density of 1.2 g/cm³, viscosity of 35 sec	20	46	31
Cementing mud, density of 1.8 g/cm³, viscosity greater than 60 sec	21	44	29

In experiments with a liquid medium, the local effect of the explosions on lead did not increase more than 20% over the experiments with air.

The slight influence of the surrounding medium on the effect of an explosion on massive lead disks placed on a stand (explosions in air and a hydrostatic pressure up to 1 kg/cm²) was demonstrated previously by the experimental study of the influence of hydrostatic pressure on the local effect of an explosion (by submerging lead targets in a hole filled with water or drilling mud). In these experiments, with constant properties of the liquid medium (water, drilling mud) and a hydrostatic pressure ranging from 75 to 300 kg/cm², very little change in the effect of the explosion was observed (see Table 20).

From experiments with massive lead targets it follows that when the charge is immersed in a medium of greater density (than air), the active part of the charge, and therefore the unidirectional impulse of the explosion, directed toward the target increases somewhat, since lateral and axial ejection of detonation products is limited by the increase in density and viscosity of the surrounding medium. This conclusion is supported as well by the results of experiments on steel disks (see Table 25).

It should be noted in conclusion that a very considerable increase in the local effect of an explosion, regardless of the medium surrounding the charge and the target face, may be accomplished by using charges possessing a directional axial cumulation. The distinctive feature of the cumulative effect is a substantial compaction of the detonation products, and thus an increase in pressure and in the energy density in these products and in the shock waves arising during the explosion.

Fig. 47. Local effect of an explosion of a cumulative charge (50 g) under a layer of drilling mud and on a lead target.

The local effect of an explosion on a lead target (under a layer of drilling mud) of a charge of liquid explosive (50 g) with a hemispherical cumulative depression, made by a glass sphere 40 mm in diameter, is shown in Fig. 47.

The possibility of using cumulative charges for explosive drilling of deep holes filled with liquid has been inadequately studied, and the matter is not discussed in the present work.

Chapter 5

THE TRANSMISSION OF DETONATIONS
BETWEEN EXPLOSIVE CHARGES

The rate of explosive drilling depends on the increment of depth at the bottom of the hole for each individual explosion (the average effectiveness of explosions) and on the frequency of setting off charges.

The explosive charges are detonated in a liquid medium during explosive drilling (in drilling muds or in water), in a space limited by the walls and bottom of the hole and at a hydrostatic pressure which, in deep holes and with the use of heavy drilling muds, approaches 1000 kg/cm^2. Under such distinctive conditions, the frequency of explosions of charges of sensitive liquid explosive mixture on the bottom of the hole is limited by the critical distance between charges according to the transmission of the detonation between charges through the liquid in the pipes. This type of phenomenon has been called mediate detonation [50].

By the critical distance r between two charges we mean the minimum distance at which the explosion of one charge does not cause explosion of another charge.

A charge that produces detonation is called active; a charge in which detonation is produced is passive.

The explosion of an active charge at the bottom of a hole may cause detonation of a neighboring passive charge in the pipe, and this may then become an active charge relative to the next charge, and so on.

The transmission of detonation between charges through a liquid has been studied by various investigators, chiefly for solid explosives, with explosions in practically an unlimited aqueous medium. These investigators have established the fact that the critical distance between charges of solid explosives is small, measuring several radii of the charge.

Experience in explosive drilling has shown that the critical distance between charges of liquid explosive of a particular composition is measured in hundreds, and in some cases thousands, of radii of the charge. In investigating the permissible frequency of explosions of liquid charges, a number of special experiments were prepared.

As is well known, some of the energy of an active charge may be transmitted to a passive charge and may cause it to detonate by the stream of detonation products, by the shock wave propagated through the medium, and also by solid particles ejected by the explosion of the active charge, or by some combined action of the indicated factors. The range of transmission of a detonation depends on many conditions [51, 52, 53, 54, 55, 56]: mass, density, detonation velocity of the active charge, properties of the passive charge determining its susceptibility to detonation, and properties of the transmitting medium. Different media apparently retard the flow of detonation products and the shock wave; the greater this retardation is, the shorter the range of transmission of mediate detonation.

The Transmission of Detonations in Air

The transmission of detonations through air has been discussed in a large number of papers. The results may be summed up in the statement that detonations in charges of high explosives are produced only at distances where the front of the shock wave and the stream of detonation products are still inseparable; in this zone the

TABLE 27. Computed Values for Density of Air at the Front of the Shock Wave and for Density of the Detonation Products for PETN and TNT

Nature of explosive charge	Density of air at front of shock wave, g/cm^3	Density of detonation products at front of projectile path, g/cm^3
PETN: density of 1.69 g/cm^3, detonation velocity of 8400 m/sec	1.283×10^{-2}	0.26
TNT: density of 1.6 g/cm^3, detonation velocity of 7000 m/sec	1.420×10^{-2}	0.25

density of the detonation products is approximately of an order that exceeds the density of air at the front of the shock wave. Thus, the production of a detonation in a passive charge by transmission through air is due chiefly to the direct action of the stream of detonation products upon the charge.

Table 27 contains computed data on the density of air at the front of a shock wave and the density of the detonation products at the front of their projectile flight, for PETN (pentaerythritol tetranitrate) and TNT [50]. Practically the same ratios are found for other explosives, including liquid explosives.

The experiments on transmission of detonations in air were conducted with spherical charges of liquid explosive weighing 20 g. For comparison, tests were made on the thoroughly studied solid explosive tetryl (density of 1.25 g/cm^3) in the form of a small pressed block weighing 10 g and having a ratio of height to diameter of 2 : 1.

A passive charge of liquid explosive was set in a hole 300 mm in diameter on a lead plate resting on a metallic ring. At a certain distance from the passive charge the active charge was suspended and initiated by an electric current.

Table 28 (the first series of experiments) shows the results of a great number of experiments with the liquid explosive mixtures 1 and 2 and with tetryl.

The Transmission of Detonations in a Practically Unlimited Liquid Medium

When the active and passive charges are separated by a solid transmitting medium, ignition of an explosion in the passive charge is possible only by the passing of the shock wave through this medium. The shock wave weakens with distance of propagation through an inert medium. If the energy of the shock wave is still sufficient at the point where it encounters a passive charge, it will generate a spontaneously accelerating chemical reaction in the explosive, as occurs by the direct action of the detonation products of an active charge through air. If the shock wave arising in the passive charge has parameters below critical values, it will pass through the charge as it did through the inert medium.

In a liquid transmitting medium a passive charge may be ignited either by the detonation products or by the underwater shock wave, which, near a charge, possesses considerably higher initial parameters than an aerial shock wave.

From Table 29, containing the computed values of initial pressure at the front of shock waves in air and in water, it may be seen that the intensity of an underwater shock wave is more than a hundred times that of an aerial shock wave [50].

The first studies were made on the transmission of detonations between charges in practically an unlimited liquid medium.* The composition and size of charges were the same as for the experiments in air.

---

*The experiments were conducted in a reservoir the size of which was sufficiently great when compared to the distance between explosive charges.

TABLE 28. Range of Transmission of Detonations between Charges of Different Explosives under Various Conditions

Series of expts	Charges — active				Charges — passive			Transmitting medium	Means of limiting medium separating charges	Range of transmission of detonations between charges, m — r_{100}	r_0
	explosive	wt., g	material and form of container (charge)	means of producing detonation	explosive	wt., g	material and form of container (charge)				
I	Mixture 1	20	Glass; sphere	KD-8	Mixture 1	20	Glass; sphere	Air	Hole, φ 300 mm in diam	0.100	0.160
		20	The same	The same	Tetryl	10	Aluminum container			0,020	0.030
		20	" "	" "	Mixture 2	20	Glass; sphere			0,010	0.025
II	Mixture 1	20	Glass; sphere	KD-8	Mixture 1	20	Glass; sphere	Water	Practically unlimited	0.500	0.90
		20	The same	The same	Tetryl	10	Aluminum container			0.010	0.02
		20	" "	" "	Mixture 2	20	Glass; sphere			0.025	0.05
III	KD-8		—	Incandescent filament	Mixture 1	5	Glass; cylinder	Water	Steel pipe, inner diam of φ 127 mm	1.5	2-4
			—	The same	Mixture 2	5	The same			0.5	1.1
				" "	Mixture 3	5	" "			0.1	
IV	Mixture 1	20	Glass; sphere	KD-8	Mixture 1	20	Glass; sphere	Water	Steel pipe, inner diam of 127 mm	40.0 [3]	—
		20	The same	The same	Mixture 4	20	The same			20.0 [3]	15.0
		20	" "	" "	Mixture 2	20	" "			8.0	9.0
		20	" "	" "	Mixture 5	20	" "			5.0	3.5
					Mixture 3	20				2.0	

* Principal properties of some liquid explosives are shown in Table 17.

** r_{100} is the limiting range corresponding to 100% production of detonations of passive charges; r_0 is the minimum range, corresponding to 100% failure to produce detonation.

*** The maximum value of r_{100} was not determined.

The active charge was immersed in water to a depth of 0.5 m and was set at some distance from the passive charge.

The comparative data obtained in the series of Expts. I and II (see Table 28) show that for charges of liquid explosive, the range of transmission of detonations in water is considerably greater than in air, especially for mixture 1; and for solid explosive (tetryl) the range of transmission in water is less than in air.

TABLE 29. Computed Values of Initial Pressure at the Front of Shock Wave in Air and in Water

Characteristics of explosive charge	Pressure at front of shock wave near charge, kg/cm^2	
	in air	in water
PETN: density of 1.69 g/cm^3, detonation velocity of 8400 m/sec	1190	195,000
TNT: density of 1.60 g/cm^3, detonation velocity of 7000 m/sec	820	136,000

The results of the experiments support the view that detonation may be produced in a passive charge of liquid explosive by transmission of the shock wave through water at a distance not attainable by direct action of the detonation products of the active charge. At the same time, the limiting range of producing detonation in a passive charge of solid explosive through water is less than for air, because of the greater retardation of the stream of detonation products by the denser transmitting medium.

Transmission of Detonations in Pipes Filled with Water

In order to produce conditions similar to those in explosive drilling, experiments were set up for transmitting detonations between charges in pipes filled with water.

At first, experiments were performed with model charges. In these experiments the capsule detonator No. 8 (KD-8) was used as the active charge, and 3 ml (about 5 g) of test explosive mixture, placed in a test tube 10 mm in diameter, was the passive charge. Both charges were set in a pipe having an inner diameter of 127 mm and being filled with water (see Table 28, series III of experiments).

The impulse from the capsule detonator (model active charge) naturally differs considerably from the impulse arising during the explosion of an actual charge containing several tens of grams of explosive; but in a pipe filled with water, even with a relatively weak impulse, detonations are transmitted to great distances (especially with mixture 1).

In order to evaluate the critical distance between charges similar to those adopted in explosive drilling, experiments were set up to transmit detonations with charges that are standard for these tests (20 g) in pipes filled with water and placed in a hole 300 mm in diameter (also filled with water).

The active charge was suspended in a column of pipe let down into the hole and was ignited by the electric detonator No. 8. A passive charge of test explosive was placed at some distance from the active charge.

The results of these experiments (series IV of experiments) show a marked increase in the range of transmission of detonations between charges in pipes filled with water, as compared with those in practically an unlimited liquid medium (series II of experiments).

The data obtained support the results of tests on model charges of liquid explosive (series III of experiments).

Of the explosives indicated in Table 28, mixtures 2, 3, and 5 permit, under the conditions of transmitting the detonations, a relatively short spacing between charges; and, in addition, they possess sufficient sensitivity to produce detonations in the charges by capsule igniters (caps.).

In contrast to the experiments in which the active and passive charges were placed in pipes, in actual explosive drilling the active charge is exploded on the bottom of the hole, at some distance from the pipe; the passive charge in the pipe is separated from the active charge on the hole bottom by a comparatively narrow channel of the so-called projectile nozzle.

A spherical shock wave, traveling out from the explosion of the active charge on the hole bottom, on entering the pipe forms a plane wave, which maintains high stability for considerable distances.

A sudden jump in pressure in the liquid within the pipe at the front of the plane shock wave depends on the parameters of the active charge, the size and position of the projectile nozzle relative to the focus of the explosion, the form of the nozzle, the dimensions of the pipe and of the hole, the form of the hole bottom and the reflecting capacity of its surface, and also the properties of the flushing medium filling the pipe and the hole.

Investigations were made with actual charges of highly sensitive liquid explosive mixture in a hole filled with drilling mud, with hydrostatic pressures ranging up to 120 kg/cm^2. In addition, corresponding observations were made during explosive drilling at depths reaching 2800 m and at hydrostatic pressures of about 350 kg/cm^2. The experiments were set up with projectile nozzles of various configurations (conical and supported), as shown in Fig. 48.

Table 30 contains the results of tests on charges weighing 50 g passed at a rate of from 2 to 22 per minute through a pipe 150 mm in diameter into a hole 300 mm in diameter, about 1000 m deep, and filled with drilling mud (specific gravity of 1.2, viscosity of 25-28 sec).

The experiments showed that when working with a supported nozzle, a passive charge in a pipe filled with drilling mud is detonated by the explosion of an active charge on the floor of the hole more than 50 m away, whereas with a conical nozzle the detonation is not transmitted so far as 16 m, which corresponds to a regime of feeding charges into the hole and of exploding them on the bottom at a rate of 22 per minute. It is clear that an increase in the critical spacing between charges when using a supported nozzle is caused by the nearness of the supporting props, resting on the bottom of the hole, to the explosion focus.

TABLE 30. Permissible Frequency of Exploding Charges of Liquid Explosive Weighing 50 g in a Hole Filled with Drilling Mud

Conical nozzle		Supported nozzle		Frequency of explosions per minute	Distance between active charge on hole bottom and place of explosion of passive charge in pipe, m
no. of expts.	number of transmissions of detonation	no. of expts.	number of transmissions of detonation		
14	None	—	—	2	—
8	The same	—	—	3	—
18	" "	—	—	4	—
35	" "	275	—	6	—
32	" "	2	2	7	52
43	" "	11	4	11	35
27	" "	2	2	22	16

During explosive drilling, especially at depths greater than 1500-2000 m, it is possible to arrange the intervals between charges fed into the column of drill pipe from the surface with a definite frequency, which should correspond to the frequency of explosions on the bottom of the hole. Under the least favorable circumstances, during explosion of a charge on the bottom of the hole, the charge nearest to the explosion may be detonated within the channel of the nozzle or immediately next to this within the drill pipe. When this happens, if the explosive drilling is conducted with unitary charges of liquid explosive, the detonation may be propagated along the entire chain of charges in the pipe, from the bottom of the hole to the feeding apparatus at the surface.

Several means are possible to prevent this from happening. One method provides the possibility of decreasing the susceptibility of a charge to detonation by changing the chemical composition of the mixture of liquid explosive. In another technique special devices may be placed above the projectile nozzle: dampers, which lower the intensity of the underwater shock wave being propagated through the pipe during explosion of a charge on the bottom of the hole.

Lastly, it is possible to place individual components (oxidizing agent and fuel), not explosive in their initial state, separately in a special envelope (capsule), subsequently mixing these into a unitary explosive charge immediately next to the bottom of the hole. In this method, throughout the entire length of the capsule supply line (this is the conduit for delivering drilling mud to the hole) from the feeding apparatus at the surface to the bottom of the hole, there moves along the pipe (or there is exploded on the bottom of the hole) but one capsule with a unitary charge at any given instant; all the other capsules moving toward the bottom contain no prepared explosive.

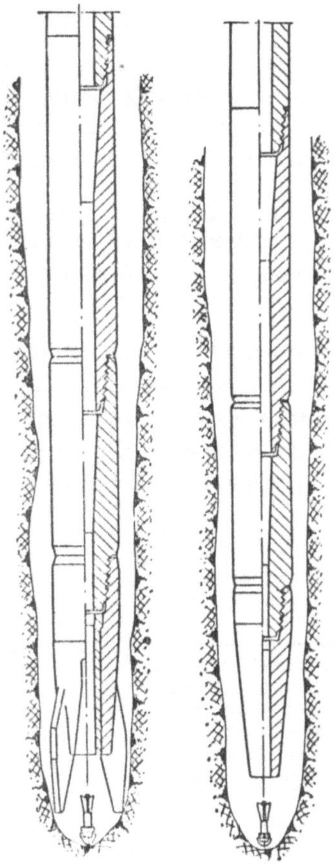

Each of the indicated methods may be used independently or in combination with each other.

We restrict ourselves here to a brief discussion of the essence of artificial damping of the shock wave passing through a pipe filled with liquid.

To diminish the intensity of the shock wave traveling through the liquid in the pipe, one may employ various devices.

Figure 49 shows a diagram of a three-stage hydraulic damper, placed in the lower part of the drill column. This construction weakens the impulse of pressure of the shock wave at each damping stage by expanding the wave front and gradually decreasing the specific energy of the explosive wave in the channels of the different segments.

It is also possible to use a damper in the form of rubber hose. It has been stated that when an underwater shock wave is propagated through such a hose its intensity will diminish because of partial loss of energy through elastic compression of the rubber.

The use of this damper in a shallow hole has demonstrated that, with an underwater explosion of a charge weighing 1.5 g, a neighboring charge weighing 7 g and placed inside a rubber hose 73 mm in diameter is not detonated at distances as short as 250 mm.

In similar experiments conducted under the same conditions, but with a metal pipe instead of a rubber hose as damper, of the same diameter as the latter, the charge was detonated on being moved as far away as 3 m.

Fig. 48. Positions of supported and conical projectile nozzles relative to the active charge of explosive in the hole.

In deeper holes the effectiveness of the damping declines; this is apparently due to the partial loss of elastic properties in the rubber; at a hydraulic pressure of about 70 kg/cm^2 this kind of damper does not fulfill its function.

Besides hydraulic and rubber dampers, it is possible to employ various pneumatic devices or a gas cap with a simple gas generator, in which the intensity of the shock wave is decreased by compressing a certain volume of air or gas. However, such a device, in principle suitable for operation in comparatively shallow holes, also loses its effectiveness quickly with increasing hydrostatic pressure and cannot be used for drilling deep holes filled with water.

Of all the examined methods, hydraulic damping should be given preference, a type in which the effectiveness must not depend on the hydrostatic pressure in the hole. For example, a five-stage damper of this type, placed directly above the projectile nozzle, permits drilling operations with a frequency of explosions up to 180 per minute (according to tests). This damper is not long (4.5 m) and the time for a capsule to pass through it, at a discharge rate of the drilling mud of 50-55 liter/sec, is little more than 0.25 sec. Therefore, in using this device, there is little probability that the assigned intervals between capsules will be such that two moving charges will be found simultaneously within the section of the damper.

However, hydraulic damping of shock waves leads to some complications, arising from secondary hydraulic loses in the system; these raise the requirements for strength in the envelopes of the capsules.

The process of explosive drilling by means of capsules carrying nonexplosive oxidizing agent and fuel separately may be conducted without a damper for the shock waves (see Chapter 7).

As experiments have shown, the narrow channels outlining the lower part of the drill column (see Fig. 58) serve the function, to a certain degree, of a hydraulic damper, since the front of a shock wave, in passing from these channels into a full section of the drill pipe, expands, and the pressure is lowered.

Observations, especially in deep holes at hydrostatic pressures in excess of 150-200 kg/cm^2, have revealed a tendency toward diminishing susceptibility of liquid explosive to mediate detonation, whereas the capacity of the explosive to ignite from a cap is preserved.

In many cases, the hydrostatic pressures of 200-300 kg/cm^2, a unitary charge of highly sensitive liquid explosive in a pipe 90 mm in diameter is not detonated at distances of several meters from the active charge (exploded on the bottom of the hole), whereas at lower hydrostatic pressures (10-30 kg/cm^2) detonations are transmitted to several tens of meters between the same explosive charges.

Experiments with charges of less sensitive explosives confirm the foregoing observations. In these experiments, when an active charge is exploded on the bottom of the hole, the pressure within the pipe, in the liquid medium surrounding the passive test charge, changes from 15 to 90 kg/cm^2. The passive charge is still detonated in the pipe at pressures of 15-17 kg/cm^2 at a distance of 35 m from the active charge on the bottom of the hole; it ceases being detonated when the pressure reaches 20-25 kg/cm^2, even when the active charge is but 20 m away.

The obtained results permit one to reject special devices for damping shock waves when using capsules bearing the individual components of liquid explosive. However, such devices may be necessary to attain the required frequency of explosions when drilling with unitary charges* of liquid explosives that have a lower susceptibility to mediate detonation.

The process of explosive drilling with no special devices for damping shock waves may be set up with practically an acceptable frequency (up to 20 explosions per minute with charges of the adopted explosive weighing about 50 g), which corresponds to a suitable regime for flushing broken rock from the hole, as established by drilling practice.

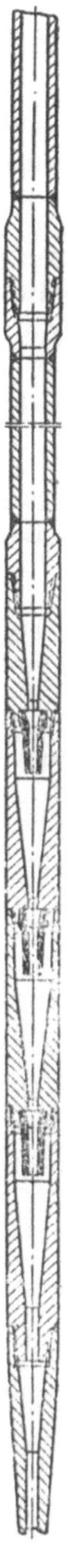

Fig. 49. Diagram of three-stage hydraulic damper with conical projectile nozzle.

The Mechanism of Igniting and Transmitting Detonations between Explosive Charges during Underwater Explosions

According to established views, the critical distance between two charges of explosive, during transmission of a detonation from one to the other through the medium separating them, varies with change in density and compressibility of the transmitting medium.

When the explosion products act directly on the passive charge, a denser transmitting medium, hampering the ejection of the detonation products, brings about a decrease in the critical distance. This relationship

---

*The use of unitary charges simplifies the construction and reduces the cost of the capsules.

is particularly clear in the example of passive charges of explosives that have low sensitivity to the transmission of mediate detonations (such as solid tetryl explosive) and that require a high igniting impulse, possible only directly next to the focus of explosion, within the zone of ejection of the detonation products.

Experimental data on the transmission of detonations between charges of solid explosives set off under various conditions (in air, in pipes filled with liquid, and in a practically unlimited liquid medium) show that the effect of these conditions of the experiment on the critical distance between charges is insignificant.

If the passive charge is highly susceptible to detonation (such as the liquid explosive used during explosive drilling), and if a comparatively weak impulse of the shock wave is sufficient to set it off (transmitted through the intervening medium), the critical distance in the transmitting medium increases markedly as compared with air. This is so because the pressure of the shock wave in the intervening liquid medium (especially when the wave is propagated through a pipe filled with the liquid) drops off much more slowly than in air, because of the slight compressibility of the liquid.

Passive charges will continue to be detonated so long as the parameters of the shock wave (on approaching the passive charge) have values exceeding some critical limit at which intense chemical reaction is still possible within the passive charge.

The critical values of the parameters of shock waves in water, i.e., waves still able (at the limit) to cause explosive transformation in charges of solid explosive, have been obtained by computation; they are shown in Table 31 [50].

Under conditions corresponding to the process of explosive drilling, the critical distance for charges of highly sensitive liquid explosive (mixture 1) may reach several thousand radii of the charge.

TABLE 31. Critical Values of Parameters of Shock Waves in the Water for Charges of Solid Explosive

Parameters of shock waves in water	Passive charge		
	mechanically stabilized RDX (cyclonite) (density of 1.40-160 g/cm^3	TNT (density of 1.30 g/cm^3)	PETN (density of 1.65 g/cm^3)
Pressure at front of wave, kg/cm^2	29,000	22,000	18,000
Velocity of wave, m/sec	3,000	2,960	2,800
Velocity of liquid at front of shock waves, m/sec	880	735	640
Temperature jump at front of wave, °C	100	80	65

The great distance that detonations may be transmitted through pipes filled with water may be explained, as indicated earlier, by the plane propagation of the shock waver under these conditions, which substantially increase the depth of the wave and the duration of its action on the passive charge.

The mechanism of setting off a detonation in a charge of explosive, independent of the mediate factor (shock wave, detonation products, solid particles ejected by the explosion), is thermal, and is similar to the mechanism discussed by Bowden and Joffe [57].

The results of Bowden's investigations are in complete agreement with the hypothesis of Berthelot, according to which the ignition of an explosion during mechanical action is caused by heating the explosion to a high temperature through conversion of mechanical energy to thermal. A thermal explosion is caused by an intense reaction due to accelerated generation of heat, greater than the elimination of the heat in the surrounding medium by convection and thermal conductivity. The mechanical energy applied to the explosive is converted to thermal energy, which is concentrated in small segments of the explosive, forming "point" sources of heat evolution and consisting of tiny gas inclusions or bubbles of air. These foci are very small (10^{-5}-10^{-3} cm in diameter), but are rather large when compared with the dimensions of molecules.

In solid explosives the excitation of an explosion (forming "point" sources of heat evolution) is possible not only from gas inclusions but also from friction on surfaces between the explosives, on surfaces of ubiquitous impurities, and on particles of the explosive itself; the excitation may also be caused by viscous heating of the explosive as the latter escapes rapidly from the space between surfaces of colliding bodies.

In liquid explosives the formation of "point" sources of heat evolution and the thermal initiation of explosions is apparently possible only by gas inclusions present in the explosive substance or introduced into it.

During explosive drilling, the shock wave, traveling through the liquid in the hole (or in the pipe), also passes through the explosive charge. As may be supposed, the mechanism of exciting detonation under such circumstances, in keeping with the data of Bowden, is associated with the development of "point" sources of heat evolution through collapse of the gas bubbles during passage of the shock wave through the explosive. Bowden has shown that the ignition of an explosive and the discharge of an explosion begin in the vapor phase, i.e., in hot gas within a bubble.

The initiation of an explosion will be achieved only if the temperature of the "hot point" proves to be sufficiently high for thermal ignition of the explosive during the time of its retention (retarded time), not exceeding the duration of action of the shock wave.

Under suitable conditions the microscopic gas inclusions lend the explosive a sensitivity sufficiently high for an explosion to be initiated by a very weak impulse.

Investigations have shown that some liquid explosive mixtures are extremely susceptible to detonation. Their ignition may be produced by an insignificant pressure at the front of the shock wave in liquid, a pressure sufficient for the formation, under certain conditions, of "hot points" in the explosive charge, which cause the explosive to ignite and then to explode. However, the sensitivity of such liquid explosives decreases with increase in initial pressure in the medium surrounding the charge, because of prior compaction, by this pressure, of the air bubbles and gas inclusions in the explosive substance.

In feeding charges into the hole during explosive drilling, a large number of charges occur in the liquid at the same time.

We may profit by the analytical determination of critical distances r_{crit} between charges as made by F. A. Baum, A. I. Gol'binder, and E. B. Kagan.

For a given distance r from the focus of an explosion, there exists the relationship

$$P = f(r, t), \tag{11}$$

where P is the pressure of the shock wave, t is the time computed from the moment of arrival of the shock-wave front to a point r distance from the focus of the explosion.

In an approximate form, for the case of an explosion in an unlimited aqueous environment, the relationship (11) may be written

$$P = P_m(r) e^{-\frac{t}{\theta}}, \tag{12}$$

where P_m is the pressure at the front of the shock wave and θ is the attenuation or damping constant [45].

The temperature in the gas bubble during its rapid collapse under the pressure of the shock wave will depend on the pressure according to a law similar to the adiabatic law. Since expression (11) gives the relationship between pressure of the shock wave and the duration of its activity, there will exist for a bubble a definite relationship between the temperature T and the time t during which the shock wave acts:

$$T = T(t). \tag{13}$$

In determining the retarded time, we may consider the gas bubble to be ignited because of the relationship, deducible from the theory of thermal explosion [50], and assuming that the ignition of an explosive (in the gas phase) occurs under adiabatic conditions:

$$\tau_{min} = \frac{e^{\frac{E}{RT_{sw}}}}{B} \frac{RT_{sw}^2}{ET_{ex}},$$ (14)

where τ_{min} is the adiabatic period of induction at a temperature of T_{sw}, E is the energy of activation, R is the gas constant, T_{sw} is the temperature of gas in the bubble compacted by the shock wave, B is a constant, and T_{ex} is the temperature, near the temperature of the explosion.

The limiting state, when the time of action of the shock wave that compresses the bubbles approaches τ_{min} at a temperature T of the compressed gas, determines r_{crit}, which, when propagation is through a pipe, is considerably greater than r_{crit} when there is spherical propagation in practically an unlimited aqueous medium.

It is convenient to determine r_{crit} graphically.

Figure 50 shows, in connection with the discussed relations (12), (13), and (14), the nature of the change in time of action of the shock wave τ_{sw} (curve 1) and the retarded time of ignition τ_{min} (curve 2) in relation to the average pressure of the shock wave P_{av} compressing the gas bubbles. The point of intersection of curves 1 and 2, where $\tau_{sw} = \tau_{min}$ at P_{av}, the value that guarantees the temperature necessary for igniting the explosive, approximately determines r_{crit} for the case of a spherical explosion.

For definite numerical values of r_{crit} it is necessary to know the thermochemical constants E and B [see equation (14)] of the explosive in the passive charge. For the explosive used in drilling these constants are not known. Special studies should be made to ascertain their values.

In speaking of the effect of hydrostatic pressure in the hole on the value of r_{crit}, it should be noted that, for a certain initial temperature in the gas bubble and for a particular excess in pressure (as compared with the hydrostatic) of the shock wave applied to the bubble, the temperature of generated heat in the bubble decreases with increase in initial (hydrostatic) pressure.

Thus, an increase in the initial pressure in the gas bubble of a liquid explosive depresses its susceptibility to mediate detonation and diminishes r_{crit}. This latter, as has been noted, agrees with experimental data on the transmission of detonations in holes at great depths, where hydrostatic pressure is relatively high.

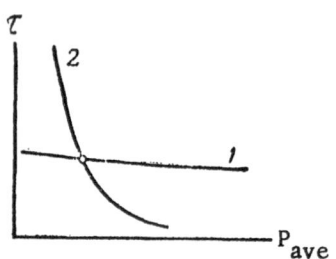

Fig. 50. The nature of the change in time τ of action of the shock wave (τ_{sw}, graph 1) and the retarded time of ignition (τ_{min}, graph 2) in relation to the average pressure of the shock wave P_{av}.

Chapter 6

REMOVAL OF LARGE ROCK FRAGMENTS FROM THE HOLE

The shattering of solid rock by impulses of high pressure is accompanied by spalling of comparatively large fragments. Under these circumstances the conditions of cleaning crushed rock from the bottom and removing it from the hole differ substantially from conditions when drilling is done with a bit. During underwater explosions some of the rock is broken off in pieces considerably larger than obtained by drilling with a bit. Furthermore, when drilling with a bit the particles of rock are broken away almost without interruption and are continuously removed by flushing. In explosive drilling, during the pauses between explosions (several seconds) the rock is not being crushed; for only a very short interval of time during the explosion (practically instantaneously) a large amount of rock is detached from the mass. The average volume of rock detached from the mass by the explosions of each charge (about 50 g) is approximately 1000 cm^3 (1 dm^3).

Depending on the state of the shaft and the depth of the hole, most of the relatively large rock fragments, weighing 30-60 g and sometimes as much as 80 g, may be removed to the surface by the discharge of drilling mud through the annular space between the pipe and the hole at a consumption of fluid (50-60 liters/sec) and velocity (up to 0.8-1.0 m/sec) as commonly employed in drilling deep oil and gas wells. A photograph of such rock particles is found in Fig. 51.

Apart from the particles of indicated size in the sludge, even larger fragments of rock have been detected in holes, the removal of which is practically impossible with the existing practice of flushing. Apparently the formation of larger rock fragments is due to hydrostatic and rock pressure, which produces stresses in the rock mass and facilitates the detachment of such fragments from the walls of the hole during disturbances from explosions or by blows of the drilling instrument during movement.

These large fragments of rock (which cannot be flushed from the hole) may shield the bottom, and when this happens a subsidiary charge of explosive is used to shatter them; thus the rate of downward growth of the hole is thus slowed.

When the spalling of such large fragments becomes excessive, technical complications may arise in the hole, substantially lowering the efficiency of the method.

We shall consider the results of experimental investigations that give us some notion of the distinctive features and of the possibility of hydraulic extraction of large rock fragments (formed during explosive drilling) from holes [58].

Model Conditions

In evaluating the possibility of making a model stand we must consider separately the scouring of sludge particles on the bottom of the hole and the removal of these particles through the space between the pipe and the hole, because the conditions and, consequently, the peculiarities of movement of particles differ for these two zones.

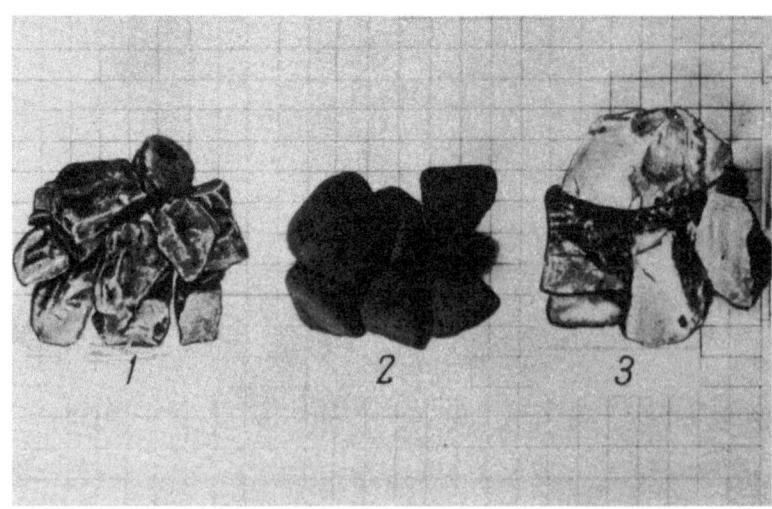

Fig. 51. Rock fragments formed during explosive drilling and flushed out of the hole. 1) Cherty limestone from a depth of about 2500 m; 2) oil-impregnated sandstone from a depth of 1650 m; 3) dolomitized limestone from a depth of 800 m.

Investigations on the model stand, in view of the complexity of the phenomena occurring in a hole, had as their objective a clarification of the qualitative picture of the process. For this reason the model conditions were determined approximately on the basis of several simplified premises.

The movement of the center of mass of a solid particle in a liquid may be expressed by the vector equation:

$$\overline{w} + \overline{q} = m\overline{a}, \tag{15}$$

where $\overline{w}$ is the vector of the resultant forces of liquid currents transporting the particle, $\overline{q}$ is the vector of immersed weight of the particle, m is the mass of the particle, and $\overline{a}$ is the vector acceleration of the particle.

By projecting (15) in the direction of the relative velocity of the particle moving in the liquid, we obtain

$$\overline{w}_v + \overline{q}_v = m\overline{a}_v. \tag{16}$$

For a spherical particle equation (16) may be rewritten in the form:

$$\frac{\gamma_1}{q}(c_1 d^2 + c_2 d)v^n + \cos\varphi \, \frac{\pi d^3}{6}(\gamma_p - \gamma_1) = \frac{\pi d^3}{6g}\gamma_p a_v, \tag{17}$$

where γ_p and γ_1 are the specific gravities of the particle and of the liquid, respectively; c_1, c_2, and n are parameters depending on the manner the liquid flows around the particle, determined by Reynolds' criterion (for structural liquids, by the generalized criterion of Reynolds[*]); d is the diameter of the particle; φ is the angle between the direction of $\overline{v}$ and the vertical; g is acceleration due to gravity; a_v is the projected acceleration of the particle in the direction of relative velocity; and v is the relative velocity of stream flow about the particle.

We shall now consider, by means of the relationship in (17), the conditions and possibilities of model construction.

Movement of Solid Particles on the Bottom of the Hole. If we preserve the average velocity of the liquid[**]in our model construction, then, at a consumption of liquid in the hole of 55 liters/sec (and a corresponding consumption on the stand at a model scale of 1 : 5, $55/5^2 = 2.2$ liters/sec) and a disposition

---

[*] The values of the parameters c_1, c_2, and n are taken from P. I. Shishchenko [59].
[**] The preservation of velocity of flow during model construction is advisable in order to decrease distortions associated with the change in coefficients in the resistance law for streamline flow about natural and model particles.

of the nozzle aperture within the working range of the bottom (250-400 mm in the hole and 50-80 mm on the stand), the liquid flow at the bottom and, apparently, at some distance from the bottom is characterized by high values of Reynolds' criterion and is turbulent, in both the actual and the model hole.

The above description applies to the use of either water or drilling mud as the drilling fluid.

When the drilling fluid flows about the particles on the bottom of the hole, the relative velocities of liquid movement may clearly prove to be rather great. This applies especially to the initial period when the particle is seized by the high-pressure stream of liquid flowing from the projectile nozzle (emission velocity of 30 m/sec at a liquid consumption of 55 liters/sec).

An approximate determination of the values of Reynolds' criterion for the streamline flow of water or drilling mud about large spherical particles (weighing 50 g and more) and about model forms of such particles indicates that during scouring of the particles in the zone about the bottom of the hole the flow will also be turbulent.

Starting with the indicated values and returning to the relationship in (17), we may compute that, in (17) $n = 2$, $c_1 \sim 0.20$ for water and ~ 0.30 for drilling mud (in the given case $c_2 = 0$ for water and for drilling mud [59].

When the particles on the bottom are being scoured, at a time when great acceleration is being imparted to them, the flow of liquid transporting the particles is resisted chiefly by inertial forces, which are considerably greater than the force of gravity. It is not difficult to convince oneself of this by examining equation (17) for particles weighing 50 g and having the above-indicated values of n, c_1, and c_2; this corresponds to turbulent flow about the particles. In addition, during this scouring the direction of movement of the particles may at some instants deviate markedly from the vertical. At such instants the gravitational forces will have values representing even a smaller fraction of the inertial forces. Therefore, for an approximate evaluation, we may disregard the force of gravity and rewrite equation (17) in the form:

$$\lambda \frac{\gamma_1 v^2}{d} = \gamma_p a_v, \tag{18}$$

where $\lambda = \frac{6c_1}{\pi}$, and a_v is the acceleration of the particle in the direction of liquid flow.

Starting from the last relationship and the necessity of observing the proportionality of the forces acting on natural and model particles (condition of dynamic similarity), we obtain

$$\frac{\gamma_{pn}}{\gamma_{pm}} \frac{a_{vn}}{a_{vm}} = \frac{d_m}{d_n} \frac{\lambda \, \varrho_{1n}}{\lambda \, \varrho_{1m}} \frac{v_n^2}{v_m^2}, \tag{19}$$

where the indices n and m signify nature and model, respectively, and l and p signify liquid and particle, respectively; ρ_1 is the density of the liquid.

Equation (19) represents the condition for model construction of scouring of spherical particles with turbulence in the liquid flowing about the partices. It was obtained approximately by selecting the following relationship between the parameters characterizing nature and the model:

$$\frac{d_n}{d_m} = \delta,$$

(where δ is the geometric scale of model construction);

$$\varrho_{1n} \approx \varrho_{1m}, \quad \mu_n \approx \mu_m, \quad \tau_{0n} \approx \tau_{0m},$$

where μ is the dynamic coefficient of viscosity and τ_0 is the limiting shears stress; $\frac{\gamma_{p,n}}{\gamma_{p,m}} = \frac{1}{\delta}$ (this is accomplished by using lead for the model particles).

In preserving the specific gravity of the particles (sludge model) for observing condition (19), the relative velocity v must be reduced proportionally to the square root of the model coefficient, i.e., in the given case by a factor of $\sqrt{5} = 2.24$. The emission velocity of the liquid from the nozzle was thus reduced by a factor of 2.24. However, the change in emission velocity is not always accompanied by a proportional change in the relative velocity v; furthermore, when the value of v changes, the coefficient c_1 may be changed.

Movement of Particles in the Space between Pipe and Wall of Hole. We shall assume that the movement of the particles in the space between pipe and hole has been established, i.e., that a definite upward velocity of the fluid corresponds to a constant upward velocity of the particles.

If there is no acceleration, the relationship (15) may be written in the form

$$w = -q \qquad (20)$$

and the regime of particle removal (during proper consumption of drilling fluid) is determined chiefly by the relative settling velocity of these particles under the effect of the force q.

In comparison with the zone about the bottom of the hole (near the mouth of the nozzle) the velocity of flow in the space between pipe and wall is considerably less, and obviously, still less than the velocity of flow about the particles. Therefore, in considering the movement of particles of different sizes, having various relative settling velocities in the fluid, one should keep in view that in going from nature to a model it is possible to change the regime of flow about the particles and, consequently, to change the character (law) of resistance.

In great measure these statements apply to the conditions when using drilling mud as the drilling fluid. As computations show, with a consumption of 55 liters/sec (or less) in an actual hole, the flow of liquid, even in a hole having the smallest diameter employed (300 mm), is not turbulent. The flow around particles, carrying the particles to the surface (including large particles), is nearly laminar both in an actual hole and in corresponding model particles in the hydraulic stand; in this process the lifting force of the flow depends on the form of the particles and on Reynolds' number.

The scouring action of the liquid and the flow about moving particles are also affected by disturbances resulting from rotation and other movements of the drill pipe; experimental data indicate that such disturbances facilitate removal of large rock particles from the hole.

Experimental Investigations

The intensity of cleaning the bottom of a hole may be characterized by the average and maximum height individual particles are raised above the bottom in a certain interval of time; it may also be characterized by the time required to clean the bottom of several groups of particles.

The intensity of removal of particles through the space between pipe and wall is indicated chiefly by the coarseness of the particles.

Figure 52 shows the diagram of a hydraulic stand. The model hole is made of transparent plastic and consists of a series of sections, which may be put together in various combinations, representing models of individual segments of an actual hole. The model (at a scale of 1 : 5) is of sections of a hole with diameter ranging from 300 to 700 mm and a total length of 12 m.

Two forms of hole bottom are modeled on the stand: a flattened form with little curvature of the surface, and a conical form.

Scouring of Large Particles On and Near the Bottom of the Hole.

For model representation of scouring of large particles on the bottom of the hole the density of the particles was varied (rock, steel, lead) and the shape and number of particles were changed, in addition to adjustments being made in emission velocity of liquid from the nozzle (from 10 to 40 m/sec) and in the form of the hole bottom. Most of the experiments were conducted with models representing the largest rock fragments, those that are not removed by such erosive action.

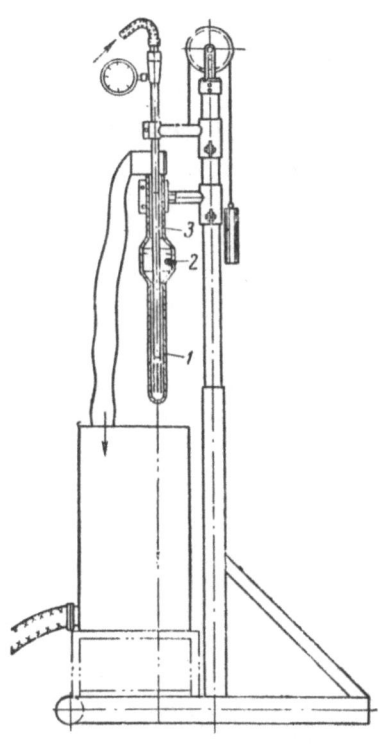

Fig. 52. Diagram of hydraulic stand. 1) Model hole; 2) model cavity (widening of hole); 3) model drilling instrument.

The nature of the movement of heavy particles on the bottom of the model hole was determined in experiments with individual particles and with groups of particles.

In the experiments with individual particles, in keeping with the previously established criterion of dimensional reproduction of the conditions, the erosion of mass resting on unit surface was approximately preserved as it is in nature. To accomplish this, particles of lead were used.

During the experiments observations were made (visually) on the nature of the movement and the height to which the particles were lifted above the bottom by the action of the stream of fluid; these observations established an average height, h_{av} and a maximum height, h_{max}, for the distance the particles were lifted after a definite interval of time (30 sec). The principal variable in these experiments was the distance l from the end of the nozzle to the hole bottom. The results of the experiments with heavy particles are shown graphically in Fig. 53. As seen from the graphs, the intensity of the stream effect on a particle, characterized by the values of h_{av} and h_{max}, depends on the distance between the nozzle and the bottom and on the shape of the bottom. These relationships are very similar for considerable variations in the consumption of fluid.

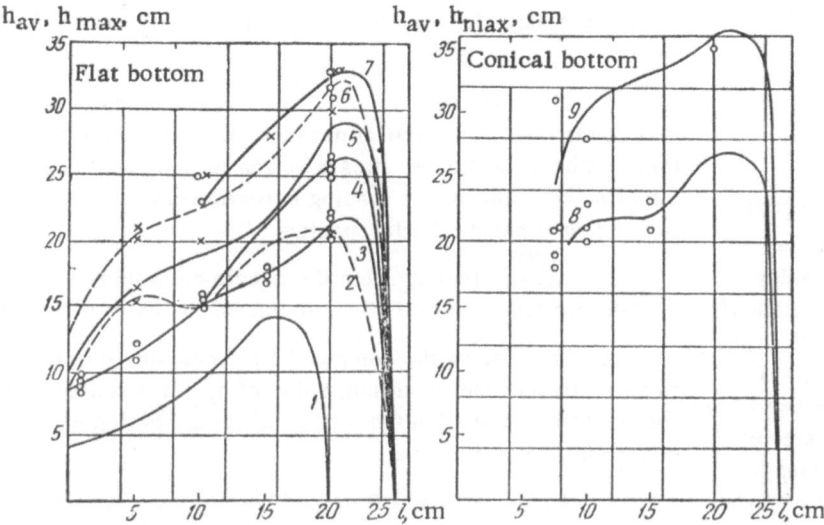

Fig. 53. Relationship between height lead spheres were raised (h_{av} and h_{max}) and distance l (in cm) from nozzle aperture to bottom, for both flat and conical forms of bottom. 1) Diam of sphere, 13.7 mm, consumption of liquid Q, 0.8 liter/sec; 2) diam = 12.5, Q = 1.4 liters/sec; 3) diam = 13.7 mm, Q = 1.4 liters/sec; 4) diam = 13.7 mm, Q = 1.7 liters/sec; 5) diam = 13.7 mm, Q = 1.4 liters/sec; 6) diam = 12.5 mm, Q = 1.4 liters/sec; 7) diam = 13.7 mm, Q = 1.7 liters/sec; 8) diam = 12.5 mm, Q = 1.4 liters/sec; 9) diam = 12.5 mm, Q = 1.4 liters/sec. Curves 1, 2, 3, 4, and 8 correspond to h_{av}; curves 5, 6, 7, and 9 correspond to h_{max}.

In the experiments with groups of particles, besides some factors studied in experiments with individual particles (distance between mouth of nozzle and bottom, and consumption of liquid), variations were made in diameter of the nozzle opening and in the form of the emitting canal from the nozzle, in the total weight and number of particles, in the specific gravity and shape of the particles (lead and steel particles of spherical and other forms), and in some other factors.

One of the indicators of intensity of scouring was the minimum time within which the bottom was cleansed of particles.

In these experiments the bottom of the model hole was flat (very little curvature). The consumption of fluid in the model, corresponding to 60 liters/sec, according to (19), with the specific gravity of model particles equal to the specific gravity of rock, was determined thus: $Q_m = \dfrac{60}{25 \cdot 2.24} = 1.07$ liters/sec.

The total weight of the particles in the experiments approximately represented the amount of sludge formed during the explosion of a single actual explosive charge. The weight of the model particles corresponded to the weight of the large fragments of rock, those exceeding 300 g.

At a model representation of liquid consumption of 85 liters/sec, the model particles corresponded to an accumulation of sludge after 8-10 explosions of actual explosive charges.

The results of the tests are shown graphically in Fig. 54; it may be seen from this graph that the time for cleaning the bottom depends on the distance between the nozzle mouth and the bottom.

The experimental investigations on scouring of heavy particles on the bottom of the model hole allow us to note some systematic relations.

1. The intensity of scouring particles, characterized by the height individual particles are lifted (Fig. 53) and by the time it takes to remove a group of particles (Fig. 54), depends on the distance between the mouth of the nozzle and the bottom of the hole. The relationships $h_{av} = f(l)$ and $h_{max} = f_1(l)$ are expressed by graphs which have a maximum point; the intensity of stream action on a particle, once having attained a maximum value, quickly falls to zero.

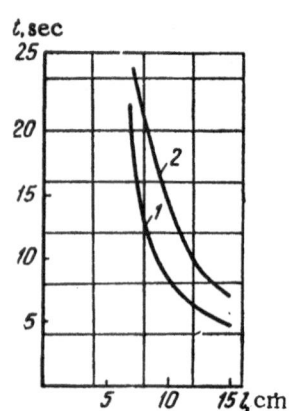

Fig. 54. Relationship between time t for cleaning the bottom and the distance l from the nozzle mouth to the bottom of the hole. 1) The amount of sludge formed from the explosion of a single explosive charge weighing 50 g; 2) the accumulation of sludge from the explosion of 8-10 charges.

2. The maximum value of h_{av} is approximately twice the value of h_{av} that corresponds to the initial distance between nozzle aperture and hole bottom.* The most intense stream action for the examined rates of fluid consumption corresponds to a spacing between nozzle aperture and bottom that is two to four times the initial spacing.

3. Streams flowing from the nozzle are able to wash out heavy particles to distances amounting to 25 diameters of the particles (see Fig. 53).

4. In changing the computed fluid consumption (for the model) from 20 to 42 liters/sec, the maximum value of h_{av} also increases almost twice, and the greatest distance (on the model) of stream action on a heavy particle increases from 1 to 1.25 m.

5. A change from a flattened bottom, concave with slight curvature, to a conical bottom leads to an increase in h_{av} and h_{max} throughout the entire tested range of distances between nozzle aperture and bottom. This fact is apparently associated with the better reflecting properties of a conical bottom, which leads to greater upward force in the stream of liquid.

6. It is noted that the shape of a particle affects the kinematics of its movement: in testing two lead particles, of equal weight, one spherical and one flat, the latter is lifted higher above the bottom than the spherical particle.

7. When the nozzle aperture is placed in an eccentric position, the stream action on the particle is more intense.

8. An increase in fluid consumption accelerates scouring of the particles on the bottom, but particles occurring directly below the jet are sometimes pressed against the bottom by the force of the flow.

The Interaction of Streams of Drilling Fluid and Particles. In other experiments the direct interaction between streams of drilling fluid, flowing from the nozzle, and the particles was observed. Attention was focused chiefly on the intersecting of streams of particles, since the effectiveness of explosions may be influenced by the shielding effect of particles on the path of a charge as it moves from the nozzle aperture to the corresponding points on the rock at the bottom of the hole.

---

*The initial distance between nozzle aperture and hole bottom amounts to about 250 mm in an actual hole, or to about 50 mm in the model hole.

On the basis of experiments conducted on the stand and of indirect data obtained during explosive drilling it has been established that particles on and near the bottom of a hole are drawn into movement by cyclic lines of flow in the liquid; these lines may repeatedly cross the jet issuing from the nozzle. This phenomenon may also be fostered by the rebounding of sludge particles after impact against the wall of the hole, against the nozzle, or against each other.

In the investigations the effect of different factors on the frequency with which particles encountered the stream issuing from the nozzle were studied; the encounter of a particle with the stream was recorded on an oscillograph by the blow of the particle on a thin metallic rod extending out of the nozzle parallel to the axis of the nozzle. Since the diameter of the rod was small (3 mm), the stream of liquid was not apparently greatly distorted by it.

From the experiments it was noted that in a given interval of observation time (30 sec), the particles crossed the possible path of the charge, and the number of these crossing increased noticeably with an increase in consumption of liquid, in distance between nozzle and bottom, and in number of particles.

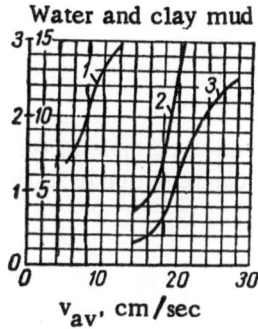

Fig. 55. Relationship of average, q_{av}, and maximum, q_{max}, weight of model sludge particles to average velocity, v_{av} of flushing liquid (water and clay mud) in the space between pipe and hole wall. 1) q_{max} (clay mud), 2) q_{max} (water, 3) q_{av} (water).

Removal of Particles from the Model Hole. The process of removing particles at various rates of fluid consumption and for different diameters of the model hole was also studied on the model stand.

For a certain size of particle being removed, corresponding to various velocities of ascending fluid in the space between pipe and wall, groups of particles were tested, representing approximately the large rock fragments (weighing from 40 to 300 g and more). The scaled consumption of flushing liquid (water and drilling mud) ranged from 43 to 100 liters/sec. The model hole on the stand permitted simulation of a cavity (widening of hole) ranging from 400 to 700 mm across.

The nature of the curves in Fig. 55 shows that when water and drilling mud are used for flushing the coarseness of particles removed is markedly increased with a comparatively small change in velocity of the ascending liquid.

The comparative data on removal of particles by water and clay mud are shown in Table 32. The clay mud exerts greater lifting force than water flowing at the same velocity. For particles of the same size, the clay mud will remove the particles at approximately one-quarter the velocity required for water.

TABLE 32. Comparative Data on Removal of Particles by Water and Clay Drilling Mud with a Scale Model of 1 : 5

Flushing liquid	Greatest diameter of model hole, mm	Consumption of flushing liquid, liters/sec	Average velocity in space between pipe and wall, cm/sec	Largest particle removed	
				nature of particle	wt., g
Water	80	1.50	37.0	Aluminum	3.0
Clay mud	80	0.35	9.0	spheres	3.0
Water	80	0.56	14.0	Sludge par-	0.3-1.0
Clay mud	140	1.80	12.6	ticles	15.0

Figure 56 shows a hydraulic stand for studying the scouring of heavy particles of rock on the bottom and for investigating the removal of these particles from the hole (on the stand) during actual drilling-mud circulation (up to 60 liters/sec).

Fig. 56. Hydraulic stand.

For observing and photographing the movement of rock particles, the upper end of the hole at the stand may be extended by means of a removable transparent section, permitting model representation of holes and cavities of various diameters.

Chapter 7

THE TECHNIQUE OF DRILLING HOLES BY EXPLOSIONS

Principal Features

In one of the variants of explosive drilling [60], when working with highly sensitive liquid explosives, the initial nonexplosive oxidizing agent and fuel are fed into the hole in special containers: capsules [61].

By using an automatic feeding arrangement [62], the capsules are filled with the chemical components and introduced at a given frequency into the circulating drilling fluid (under pressure), by which they are carried down the column of pipe to the projectile nozzle, where they gain acceleration and are detonated, at the same frequency, when they come into contact with the bottom of the hole.

Fragments of rock broken away from the mass by the explosions, as with ordinary mechanical drilling, are removed by circulating drilling fluids through the annular space between the column of pipe and the hole wall.

In the final interval of movement of the capsule to the bottom of the hole, a charge of liquid explosive mixture is prepared automatically from the nonexplosive chemical components.

The capsule passes through this interval in the time necessary to form the liquid explosive mixture.

The time required to mix the initial components to produce the desired size of charge is about 1.5 sec. Consequently, a frequency up to 40 explosions per minute may be achieved by this variant of explosive drilling.

It is also possible to use such oxidizing agents and fuels that are not in themselves explosive to form unitary charges of explosive mixture directly in the feeding mechanism, if the susceptibility of this mixture to transmission of detonations between charges is such that an acceptable frequency of explosions is possible. For this technique it is possible to employ capsules of simpler construction and to simplify the instrument by which the explosive drilling is to be effected.

A nozzle is designed for flushing the bottom of the hole of rock fragments broken by the explosions and for accelerating the movement of the capsules toward the bottom with minimum scattering.

The maximum pressure at the front of the shock wave, which is propagated through the liquid during the explosion, reaches 200,000 kg/cm^2 immediately next to the charge. Away from the focus of the explosion this pressure falls off rapidly, not exceeding 2000 kg/cm^2 at a distance of 200 mm for a charge of liquid explosive weighing 50 g, which is little more than one-fifth the limiting strength of the steel from which the nozzle is made. Nozzles are constructed with smooth graduations or transitions where stresses are concentrated, streamlining the passage of shock waves around them.

These nozzles are variously designed. Their protection from the direct action of an explosion is achieved by maintaining a safe distance between the nozzle and the explosion focus on the bottom of the hole.

One type of projectile nozzle has props, the projection of which exceeds the height of the charge and the full length of the capsule. Such a nozzle rests on the bottom during explosive drilling and is therefore called a supported nozzle.

Another type of nozzle, called conical, must be held at least 200 mm from the focus of the explosion for the charges of explosive used in these studies, and no more than 400 m, to prevent undue scattering of the capsules over the bottom.

The minimum distance is controlled by designing capsules with spacing (or distance) guides; the maximum distance is controlled by acoustical methods.

The Explosive Capsule

The designated capsule is for transporting the explosive charge to the bottom of the hole and to initiate detonation in it; the work of the explosion will be reflected in shattering of the rock and in formation of a hole.

We have previously considered the basic parameters of underwater explosions and of the zones of influence of the explosions—in solids, in models (Chapter 2), and in rocks in their natural occurrence (Chapter 3). In these chapters we studied the factors that modified the effectiveness of explosions and the diameter of a hole.

During the experimental investigation of factors effecting the operation of a charge, tests were made on physical properties of explosives, on the production of detonations, and on the completeness of explosions under high hydrostatic pressures (Chapter 4). The investigation of the properties of explosives under these conditions was conducted also in connection with the study of transmission of detonations through the liquid medium surrounding the charge (Chapter 5).

The results of the investigations served as the basis for choosing explosives and for determining size and shape of charge.

In working out the design for an explosive capsule, consideration was given to the necessity of stabilizing the movement of the capsule as it is ejected from the nozzle, since excessive scattering of the capsules about the bottom of the hole leads to lower effectiveness per explosion, causes cavity development in the shaft of the hole, and increases the diameter of the hole. In addition, the capsules were given special properties (spacing guides), conditioned by the technological requirements of drilling the hole.

A capsule (Fig. 57) should have a strength sufficient to preserve the explosive charge in its initially prepared form till the instant of detonation (when it strikes the bottom). To avoid removal of the unitary charges to the surface of the ground, the body of the discharged capsule should be destroyed by the blow against the bottom of the hole, and the liquid explosive mixture should pass harmlessly into the circulating drilling fluid.

The explosive capsules should be safe in handling. To satisfy this requirement, the charges of liquid explosive are prepared automatically in the hole or in a feeding device during the process of explosive drilling; they are made of components not explosive in their uncombined state.

When these requirements are varied because of the use of charges of solid explosives, great care must be given to guaranteeing the safety factors in the construction of the capsules and in providing for proper transport, storage, and use of the charges.

The capsules are introduced through a feeding device where, after being filled with the chemical components in one of the automatic operations, their final assemblage is completed and they are sent into the hole.

The dimensions of the inner cavities of the capsule are determined by the size of the charge, and the relative volumes of the cavities depend on the properties of the chemical components of the explosive; these components are generally used in quantities near their stoichiometric proportions.

To relieve the capsule from the dynamic pressure of the pumps forcing the drilling fluid into the hole and from hydrostatic pressure, which reaches hundreds of atmospheres in deep holes, a simple device is provided that connects the inner cavities of the capsule with the surrounding liquid medium.

As already pointed out, the detonation of liquid explosives of a certain composition is achieved by a thermal impulse from safe igniter capsules.

The mechanism by which the detonators in the capsules work is considered in relation to the inertial forces acting on them because of deceleration caused by movement of the capsules along a path with channels of variable cross-sectional areas.

The tail of the capsule ensures oriented movement of the capsule through the pipe, stabilizes it as it issues from the projectile nozzle, and guards the detonator, not permitting it to perform its function except under proper conditions.

If the capsule is detonated in the channel of the nozzle, too near the bottom of the hole, then no matter how strong the steel, regardless of the thickness of the nozzle walls, the metal will be shattered after several explosions. The same thing may happen when several explosions are detonated immediately next to the nozzle, causing collision between the capsule, as it emerges from the nozzle, and large rock fragments that move in random paths in the circulating drilling fluid.

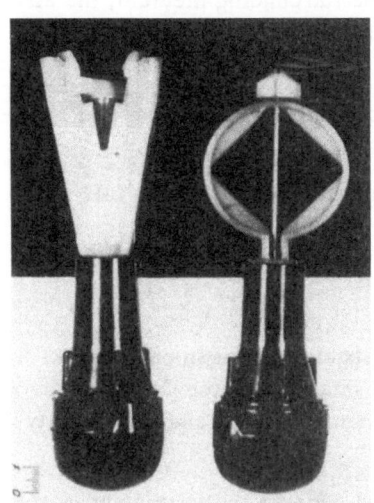

Fig. 57. Outer view of capsules with spacing guides (capacity of capsules, 33 ml).

The detonator preserves the nozzle from shattering by explosions, since it operates by the blow of the capsule against the bottom of the hole only when it is a definite, safe distance from the nozzle.

If the capsule, having reached the bottom of the hole, has not completely emerged from the throat of the nozzle, which is too near the explosive charge and susceptible to damage by the explosion, the casing of the capsule is destroyed and is ejected from the nozzle under the pressure of the flushing liquid, and the explosive mixture is washed away and rendered harmless in the stream of fluid.

When the distance between the nozzle and the bottom exceeds the height of the capsule, then the explosive mixture prepared during the passage of the capsule to the bottom of the hole is detonated by the impact.

An operator adjusts the instrument by hand (or the adjustment is made automatically), in proportion to the increment of the hole, when the sound of the explosion is heard or when a signaling device indicates the capsule has been detonated on the bottom. If the signals cease because the nozzle gets too close to the bottom, the operator increases the distance between nozzle and bottom and restores the operation of explosive drilling. By periodic "sounding" of the bottom, i.e., by artificially stopping the explosions, an excessive distance between nozzle and bottom (more than 400 mm) is avoided.

The discussed capsule was designed for components that are combined immediately next to the bottom of the hole to form a highly sensitive liquid explosive mixture. Such a mixture cannot be used for explosive drilling in a unitary form because of mediate detonation between charges; such detonation is transmitted great distances through the liquid medium.

Certain liquid explosives possess less susceptibility to the transmission of detonation and may be used for explosive drilling by means of unitary charges.

The nature of the reaction between the explosive and water has proved to have a definite influence on the design of the capsules. Some liquid explosives are nonhygroscopic, are insoluble in water, and do not require sealing; but others, mixing with water, lose their properties.

This feature of explosives requires, if not complete sealing, at least strict prevention of access of water to the charge. The possibility of regulating the rate of decomposition of such explosives in the presence of water may be used for spontaneous destruction of unitary charges, deactivating the capsules in the time required for their removal by the drilling fluid from the hole to the earth's surface.

In one of the simplest plans for spacing the capsules, the capsules are inverted as they emerge from the nozzle.

A capsule consisting of an envelope with the simplest type of inertial igniter moves along the pipe and into the nozzle tail end foremost. If the capsule strikes the bottom in this position no explosion will occur. On emerging from the nozzle the capsule is rotated 180°, and, set in the proper position, it is exploded on impact with the bottom.

The moment necessary to rotate the capsule as it leaves the nozzle is effected by selecting materials of different densities and by using hydrodynamic forces through suitable design of the head and the tail of the capsule.

For simplicity and economy in the manufacture of the capsules, the possibility of direct initiation by air of an explosion in the explosive charge (without any intermediate igniter or detonator) deserves attention.*

When the casing of a capsule as described strikes against the bottom of the hole with a certain velocity, it is deformed, and the seal is destroyed. Because of the excessive pressure in the surrounding medium, the air bubble contained in the capsule at atmospheric pressure collapses with great velocity, and an explosion is initiated by the heat (in combination with the shock).

The air bubble in the capsule (it is the initiator) ever assumes a higher position, and in this way the greatest possible local effect of the explosion is obtained.

Under ordinary conditions (at atmospheric pressure), initiation by air ensures safety in handling fully equipped capsules.

The possibility of initiating explosions in the explosive charges by means of air has been confirmed experimentally at various hydrostatic pressures.

Difficulties are encountered when pneumatic initiation is combined with spacing arrangements on the capsule. Therefore, the use of such capsules is suitable for explosive drilling of small-diameter holes with smaller explosive charges, because under such circumstances, with the development of sludge and the stability of the projectile nozzle, the distance factor is of less significance.

By using capsules with spacing guides, shattering of the nozzle is prevented because the instrument is restrained from approaching too close to the focus of explosion.

However, such capsules do not guarantee automatic control of the position of the nozzle above the bottom of the hole at distances exceeding the full height of the capsule. Therefore, when drilling at great depths, when the length of the column of pipe (because of its weight) and technological complications make it difficult to control the maximum distance between nozzle and bottom, explosive drilling with capsules having spacing guides is expediently combined with acoustical-sounding control (this method will be discussed separately).

The Equipping-Feeding Apparatus

During explosive drilling important functions are filled by the apparatus, essential for this type of drilling, that equips the capsule with the chemical components individually, not explosive in their separate state, and that feeds these charges into the hole and to the bottom at a rate corresponding to the required frequency of explosions. The assemblage of automatic apparatus that accomplishes these tasks is called the equipping-feeding apparatus.

The Instrument

The conduit through which the drilling fluid and the capsules are forced into the hole from the equipping-feeding apparatus and thence to the projectile nozzle is smooth and round with smooth graduations and joints, designed with proper consideration to the dimensions of the capsule and to its velocity in the various segments of the passage.

Nipples and sleeve coupling are attached to the ends of the drill pipes, by screwing on (external threads) or by joint welding; these connect the individual pipes in the drill column (of equivalent cross section).

As the hole deepens the column is lengthened by adding single pipes (11-12 m long), and during raising and lowering operations two or three connected pipes are handled as a unit, depending on the height of the drilling derrick. Fundamentally the column of drill pipe is the same as that used for drilling with a bit.

During explosive drilling the first unit to reach the bottom is a special device differing substantially from parts used in other methods of drilling; this unit is a shaped part of the drill column, 24-25 m long (Fig. 58). All

_____
*Proposed by V. N. Grinblat and E. B. Kagan.

96

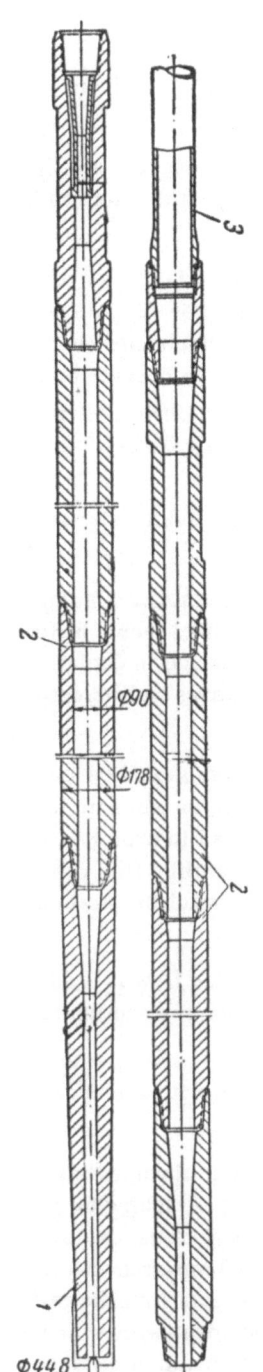

Fig. 58. The shaped part of
the drill column. 1) Projectile
nozzle; 2) weighted drill pipe;
3) drill pipe.

gradations in size in the shaped part of the column are effected by intake
and discharge cones, thus securing uniformly accelerating (and uniformly de-
celerating) movement of the capsule to the projectile nozzle.

Over a considerable part of the journey of the capsule along the heavy
drill pipe to the bottom of the hole, the oxidizing agent and the fuel are
mixed and a charge of liquid explosive is formed.

To obtain a liquid explosive mixture from nonexplosive components,
the period of movement of the capsule through this segment of its journey
should be at least 1.5 sec; for a given consumption of drilling fluid and a particular cross section of pipe, this time is determined by the length of the drill
column.

When detonations are transmitted between charges, at the time a capsule is exploded on the bottom of the hole, the shaped segment of the column should be free of any capsule. To ensure this the capsules are fed into
the hole at a frequency, the time interval of which is known to be greater
than the time required for the movement of a capsule through the shaped segment of the column. In practice, when the time required for the movement
of a capsule through the shaped part of the column is about two seconds, the
pause between instants of feeding capsules into the hole should be somewhat
greater, but the frequency becomes correspondingly less.

Besides the indicated function, the shaped part of the drill column
serves as a heavy base, facilitating better centering of the instrument in the
hole.

During operations with unitary charges of explosives (single-chambered
capsules), the nozzle is eliminated from the assemblage of the shaped part of
the drill column.

The Projectile Nozzle

The most important part of the shaped segment of the column is the
projectile nozzle. This nozzle is designed for flushing the bottom of the rock
shattered by the explosions, for accelerating the passage of the capsules, and
for directing the capsules toward the bottom of the hole with a minimum of
scattering.

A conical or a supported projectile nozzle may be used, depending on
the method of controlling the distance between the nozzle and the bottom of
the hole.

With a conical nozzle, the process of explosive drilling is carried out
with the nozzle held slightly above the bottom of the hole, and, consequently, held a corresponding distance from the focus of the explosion.

The conical nozzle (Fig. 59) has minimum cross-sectional dimensions
and a streamlined form, simulating the configuration of the near-bottom segment of the hole, as this zone develops by the passage of the explosive process through most rocks. A conical nozzle of proper form reduces to a minimum the possibility of the nozzle becoming jammed in the hole, and the
presence of smooth connections at places where stresses are concentrated
makes the conical nozzle more stable than the supported nozzle (Figs. 60, 61).

When necessary, the axial load on a conical nozzle may be reduced to 15-20 tons. When this is done the
arc at the apex of the nozzle forms a gap, sufficient for normal flow of the drilling fluid. A supported projectile
nozzle stands on the bottom at the time of the explosion, bearing a definite axial load; the supporting props,
being near the explosive charge, are subjected to considerably greater stresses than a conical nozzle. Despite
the comparatively high stability of a conical nozzle and the simplicity of its design, under certain conditions it

may prove expedient to use supported nozzles as well; these latter permit one to maintain the required distance above the bottom for work with the capsules of simplest design, those having no spacing guides for the explosions and containing no acoustical apparatus.

The forces acting during explosions on the bottom of the hole tend to rupture the supported projectile nozzle along the joins or seams.

Experience has shown that when the connections between the props and the sleeve of the nozzle are rigid, after several hundred explosions of 50-g explosive charges fractures begin to develop rapidly on the front surface of the nozzle, as shown in Fig. 60.

Figure 61 shows variant designs for a projectile nozzle with spring supports, in which the face (sleeve) part of the nozzle is disconnected from the supports. In one of the nozzles the release is achieved by deep arcuate indentations, in the other by the considerably elongated elastic elements of the design, for which slits have been provided, grading into "discharging" openings. By giving the nozzle props some mobility it has become possible to increase the number of explosions to 5000-6000 between repairs on the props, as against several hundred explosions when using a nozzle with rigid props.

In order to increase the stability of the supported nozzles several general conditions should be adhered to:

a) The contour of the props for the projectile nozzle, facing the channel, and the path of the entrance into the channel should be formed on a curve, with the possibility of uninterrupted change in curvature; this provision decreases retardation of the shock waves and lessens any local rise in pressure.

b) The cross-sectional dimensions of the props should be minimal, as much as this is possible while maintaining the strength of the nozzle proper to the technological features involved in explosive drilling; the minimal dimensions should improve streamline flow of the shock wave about the props and should reduce the resulting strain associated with this wave.

c) In order to improve conditions of streamline flow it is advisable to shape the props so that the greatest curvature of a section faces the explosion focus.

d) The props should be given some mobility, with due consideration to the previously indicated conditions.

Breakage in the nozzle by the direct action of shock waves during explosions of charges of the adopted explosive (weighing 50 g) is possible only when the nozzle is very near the focus of the explosions. According to experimental data, a distance of 100 mm from the nearest point of the charge is enough to protect the nozzle from the destructive action of the explosions. This points up the advantages to the stability of a conical nozzle, which is suspended the required distance above the bottom of the hole, where the charge is exploded. If this distance is strictly observed, the pressure created by the explosions at the points in the liquid touching the face of the nozzle produces no stresses that are dangerous to the strength of the material and no fatigue phenomena in the metal.

However, when drilling deep holes, because of the ever-possible technological complications, a controlled distance between nozzle and bottom of hole is disturbed, and, furthermore, it is possible that capsules with defective detonators will fire at the contact with the nozzle or immediately next to it. Therefore, in practice, the conical

Fig. 59. Conical projectile nozzle (discharge diameter of the nozzle aperture of 45 mm).

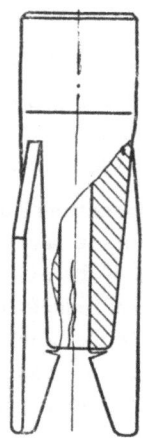

Fig. 60. Supported projectile nozzle. Cracking is shown along the seam of the nozzle and incipient rupture at the base of the props.

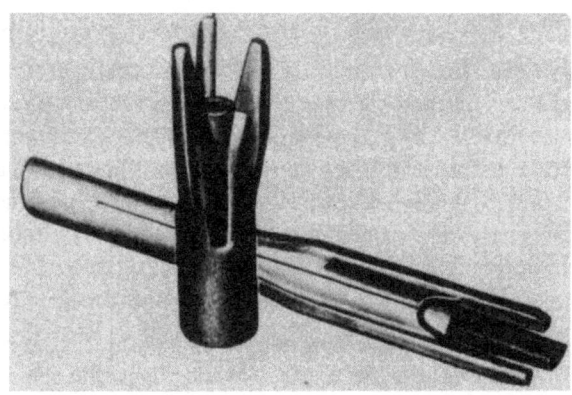

Fig. 61. Projectile nozzles with spring supports (discharge diameter of nozzle tube, 45 mm).

projectile nozzle is constantly being damaged, and its condition should be checked after approximately 10,000-15,000 explosions.

An external view of the face of a conical nozzle after 12,000 explosions, the instrument having passed through more than 100 m of hard rock in a deep hole, is shown in Fig. 62. High-strength alloy chrome-nickel and chrome-nickel-molybdenum steels have been used in the manufacture of the projectile nozzles; these steels have high resilience. The mechanical properties of the steels for the nozzles are shown in Table 33.

TABLE 33. Mechanical Properties of Steels for Manufacturing Projectile Nozzles

Brand of steel	Mechanical properties					
	yield point, kg/mm^2	ultimate tensile strength, kg/mm^2	elongation, %	cross-sectional contradiction, %	resilience, kg-m/sec	Brinell hardness, diam of impression
OKhN3M	85	90	15	52	10	3.9
EI-579:						
longitudinal samples	65	74	17	70	20	3.9
transverse samples	55	70	16	62	9	4.0
OKhN1M:						
longitudinal samples	64	78	19	68	17	3.9
transverse samples	63	78	17	57	11	3.9

Method and Apparatus of Control

The principal parameters for explosive drilling of deep holes depend in great measure on the stability of the projectile nozzle (the conical nozzle is implied), as measured by the number of explosions it withstands.

As already pointed out, the "potential" stability of a nozzle may be rather great under certain conditions; but if the nozzle is held too close to the focus of explosion on the bottom of the hole, it begins to undergo premature destruction. The necessary stability in a nozzle for operation with the comparatively large charges of the explosive employed (weighing 50 g) may be achieved only by reliable control of the distance between nozzle and bottom of hole.

The simplest, but not yet perfected, method of control is considered to be periodic "probing" of the bottom, in the pause between groups of explosions, with the projectile nozzle and subsequent re-establishment of the initial spacing (safe for the nozzle) between nozzle and bottom. The periodicity of this process is attained by a definite regime of feeding capsules into the hole, in keeping with the manipulation of the drilling instrument and effected according to any given program by means of special automatic apparatus.

Apart from the great loss of unproductive time (nearly 50%) resulting from group feeding of capsules and periodic "probing" of the bottom, this method of control has still one other disadvantage: with growth of the hole the stretching of the column of drill pipe increases because of the added weight, and this fact lowers the precision of determining the actual distance between nozzle and hole bottom. This conditions is aggravated by friction of the pipe against the wall, by the adherence of the pipe to the clay crust, and also by sticking of the column to projections and irregularities in the shaft of the hole. Under such conditions it is easy to set the nozzle too far from the bottom, which may lead to excessive scattering of the capsules, poorer flushing of the bottom, less effectiveness of the explosions, and the formation of cavities in the sides of the hole.

Fig. 62. External view of the face (operating) part of a conical projectile nozzle after 12,000 explosions.

The use of capsules with spacing guides prevents damage to the nozzle when the nozzle is too near the focus of explosion or when a capsule strikes a large rock fragment on emerging from the nozzle (such fragments may be moving haphazardly in the circulating stream of drilling fluid). However, a capsule with a spacing guide does not ensure control of the distance between nozzle and bottom at distances that exceed the height of the capsule. Therefore, during explosive drilling at great depth, especially when conditions are complex, distance control between nozzle and bottom by means solely of spacing guides on the capsules must prove to be inadequate.

The development of a control for position of the projectile nozzle, without touching the nozzle to the bottom of the hole, is a complex problem.

Actually, at hole depths, and length of drill pipe, of approximately 2000 m, it is necessary to set the distance between nozzle and bottom with a precision of 200 m, or 0.01% of the length of the column; the actual distance should be between 200 and 400 mm. This precision is made difficult because of the lack of a reliable channel connecting the bottom of the hole with the earth's surface. The use of a wire connection inside the pipe along the path followed by the capsules is not acceptable, and an outer pipe is impossible if the channel is to be protected from mechanical damage.

The use of radio location is impossible because of the impermeability of water to radio waves.

A magnetic system of communication by using the column of drill pipe as a magnetic conductor is difficult because of the great number of breaks in the magnetic change (at the joints), because of the great length (with small cross section), and because of the low magnetic permeability of the pipe steel and, consequently, the great magnetic reluctance of the column. Furthermore, this system exhibits great parasitic conductivity between contact surfaces of metals in the drill column, the conductor, and the special column. It is clear that the most acceptable solution is the use of acoustical oscillations arising during movement of the capsule and its explosion on the bottom of the hole.

Practical interest is being shown in the column of pipe and the column of drilling fluid contained within it under pressure as a possible communication channel.

Three types of mechanical oscillations may be sent through the body of the column of pipe: longitudinal, transverse, and torsional. Torsional and transverse oscillations cannot be used, because it is difficult to find an actual method for producing them in the indicated circumstance. Longitudinal waves will pass through the body of the pipe with great loss because they die out in the metal by reflection of acoustical waves at the joints, and also because of friction between the column, the walls of the hole, and the drilling fluid.

There is considerable interest in the liquid filling the drill pipe as a communicating channel. The absorption coefficient of acoustical waves in liquid depends on the frequency of vibration, the propagation velocity of the waves in the liquid, the kinematic viscosity of the liquid, and the presence of air bubbles.

An experiment in the use of hydro-acoustical communication has shown that, if we take into account the comparatively small extent of the column of liquid and the relatively low frequency of the oscillations arising during an explosion, this connecting channel may prove to be the most reliable for transmitting signals from the bottom of the hole to the earth's surface.

In order to determine the controlled distance by sounds emanating from the bottom of the hole, it is necessary to obtain at least two signals, one of which must come from the bottom and the other from some variable distance H which includes the distance h between the bottom and the nozzle.

We shall examine some possible techniques of measuring the distance between bottom and nozzle.*

The first design (Fig. 63) specifies that the capsules will be used as the source of oscillations. The passage of a capsule through a window is accompanied by an impact that is recorded by an instrument. The following signal, the explosion of the capsule on the bottom, is also recorded. Knowing the velocity v and time of travel of the capsule t, we may determine the distance h;

$$h = vt.$$

In determining the distance between the nozzle and the bottom of the hole according to the second plan (Fig. 64), a sound emitter (magnetostrictional, electrodynamic, or other type) sends short acoustical signals into the column of pipe. If the generating signal coincides in phase with the reflected signal, the Lissajous figure obtained on the screen of a cathode oscillograph forms a straight line, inclined at an angle of 45° to the vertical axis. The wavelength of the signal obtained is $\lambda = 2h_{opt}$ ($\lambda = 4h_{opt}$, etc.). The apparatus is set with the drill-pipe column in a position corresponding to the initial position before any charges have been exploded; this initial position places the nozzle at the optimum distance (h_{opt}) from the bottom of the hole.

The value of h_{opt} is maintained by vertical adjustment of the column of pipe till the Lissajous figure assumes its initial form (an inclined line).

The principal disadvantage of this system is the difficulty of making measurements, because of the low intensity of the reflected signal and because of multiple reflections from capsules that are inside the column of pipe.

In using the third design (Fig. 65), a sound emitter sends short (attenuating or with constant amplitude) acoustical signals into the column. Two reflected signals are received (from points A and B). From the lag of the second reflection (from the bottom) the distance H is determined (from the time t). The difficulty of using this system involves the very small reflecting surface at the point A and, consequently, insufficient intensity of the reflected signal. In addition, the sound receiver will also receive signals, in this case, reflected from capsules moving along the column of the drill pipe.

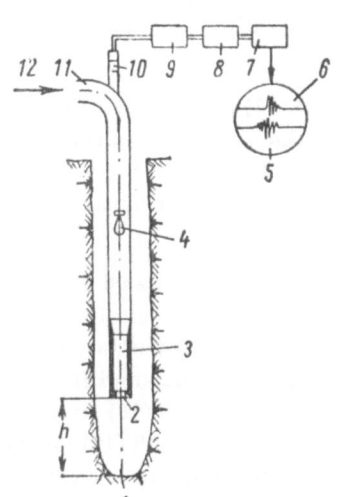

Fig. 63. Control of distance between bottom of hole and projectile nozzle (first design). 1) Site of explosion; 2) window; 3) discharge channel; 4) capsule; 5) oscillogram of impact of capsule when passing through window; 6) oscillogram of explosion; 7) oscillograph; 8) band filter; 9) amplifier; 10) sound receiver; 11) pipes; 12) drilling fluid.

From a study of the range of sound vibrations in the liquid filling the column of pipe, and from the frequency characteristics of the vibrations arising during an explosion of a capsule on the bottom of the hole, the first system (see Fig. 63) has been adopted for working out a method of controlling the distance between nozzle and bottom of hole. According to this design, the distance between nozzle and bottom is maintained automatically by two pressure impulses recorded at the earth's surface. The first impulse of pressure coincides with the entrance of a capsule into the constricted part of the nozzle, the second with the explosion of this capsule on the bottom of the hole. Both signals travel with the velocity of sound through the column of liquid and are received at the surface by piezoelectric vibratory receivers of special design.

*Developed jointly with I. I. Sud.

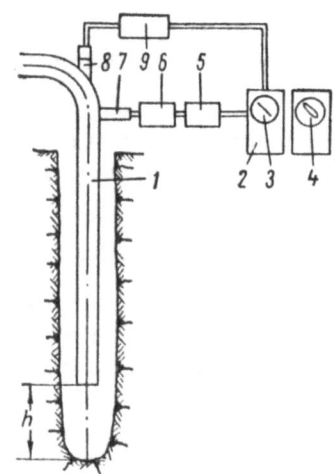

Fig. 64. Control of distance between bottom and nozzle (second design). 1) Pipes; 2) cathode oscillograph; 3) Lissajous figure at optimum distance (h_{opt}); 4) Lissajous figure at distance h, deviating from optimum; 5) band filter; 6) amplifier; 7) sound receiver; 8) sound emitter; 9) generator.

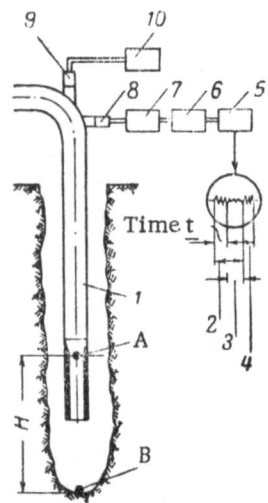

Fig. 65. Control of distance between bottom and nozzle (third design). 1) Pipes; 2) reflected wave from point A; 3) interference; 4) reflected wave from bottom (point B); 5) oscillograph; 6) band filter; 7) amplifier; 8) sound receiver; 9) sound emitter; 10) generator.

The vibratory receiver has two electromechanical converters, consisting of disks of barium titanate incorporated in a differential scheme for compensation of the noise and for distinguishing a useful signal transmitted through the liquid (and through the metal of the drill pipe).

Each converter is glued to a separate steel membrane forming the race of an oil chamber. The inside of this chamber communicates with the circulating drilling fluid. Thus, when an impulse of pressure traveling through the liquid is received, the membranes are bent toward the opposite side, which produces an effect at the outlet of the vibratory receiver of a total signal from the two converters.

The oscillation of the noise arising during drilling operations acts unilaterally on a converter, and because of counter connections, the signal from the noise is zero at the outlet of the vibratory receiver. The distance between nozzle and bottom is determined by the arrival times of the first and second pressure impulses at the earth's surface and by the velocity of the capsule. These same impulses are used for communications from the automatic feeding device to the instrument when there is an increase in the pause between pulses due to the increment in depth of hole produced by the explosions.

A circuit diagram of the arrangement for controlling the position of the projectile nozzle relative to the bottom of the hole during explosive drilling is shown in Fig. 66.*

An impulse from the vibratory receiver (1) is recorded by a recording head (3) on a drum (2), the surface of which is covered with magnetic material. To extend the dynamic range of the record, the amplitude-modulated signal is converted to a frequency-modulated signal by means of a modulator (4).

The drum is rotated by a synchronous motor (6) through a reducer (5). The record is produced by an attachment (7) and transmitted to the intake of a demodulator (8), where the frequency-modulated signal is converted back to an amplitude-modulated signal. The amplified and filtered signal is sent to a cathode-ray tube (9). Contact (10) starts horizontal scanning by the cathode-ray tube of an oscillograph (11) synchronously with the rotation of the drum. The sawtooth voltage actuates a scanning unit (12). An amplitude selector is placed at the outlet of the demodulator. On arrival of a signal from an ejecting capsule, with an amplitude exceeding the arresting level, relays (14) and (15) switch off the recording head (3) and the erasing attachment (16). In this way the drum retains records of the ejection impulse, the explosion, and the oscillations preceding and following these impulses. At each rotation of the drum the process is reproduced on the screen of the tube of the cathode oscillograph, which has a long afterglow.

Thus, one sees an immobile picture on the screen of the oscillograph, showing the process recorded on the drum.

The scale of the grid on the tube screen is calibrated in units of length. Computation of the distance between two impulses on the screen therefore gives immediately the distance between nozzle and bottom of hole.

---

*Developed by the Bureau of Design for Petroleum Instrument Construction.

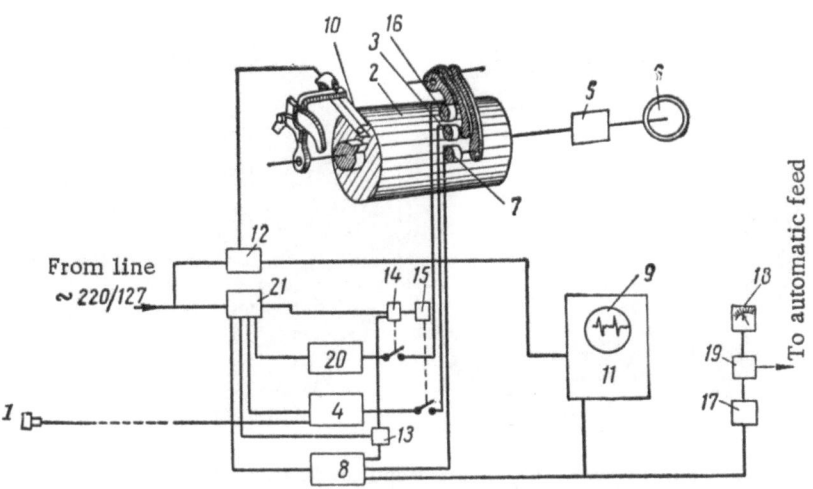

Fig. 66. Circuit diagram of arrangement for control of distance between projectile nozzle and focus of explosion on bottom of hole. 1) Vibratory receiver; 2) drum; 3) recording attachment; 4) modulator; 5) reducer; 6) synchronous motor; 7) reproducing attachment; 8) demodulator; 9) cathode ray tube; 10) contact; 11) cathode oscillograph; 12) scanning unit; 13) amplitude selector; 14 and 15) relays; 16 erasing attachment; 17) arresting device; 18) calibrator; 19) comparison unit; 20) voltage regulator; 21) unit for stabilizing feed.

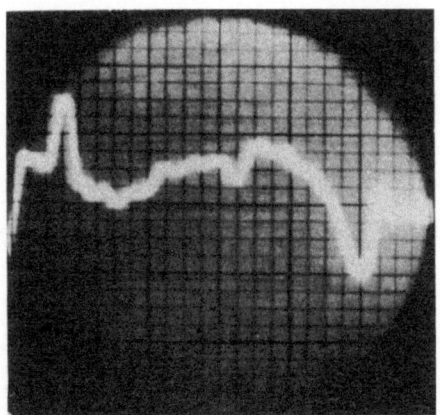

Fig. 67. Oscillogram of pressure impulses corresponding to the entry of a capsule into the channel of the projectile nozzle and to the detonation of the capsule on the bottom of the hole at a distance of 200 mm between nozzle and bottom (depth of hole, 1000 m).

Figure 67 shows an oscillogram on which are recorded pressure impulses that correspond to entry of a capsule into the channel of the projectile nozzle and to detonation of the capsule on the bottom of the hole at a depth of 1000 m and with a distance of 200 m between nozzle and hole bottom.

An analysis of a great number of oscillograms has shown that the maintenance and control of a distance of 200 m between nozzle and bottom may be attained with a maximum error of 12.5%, i.e., within ±25 mm

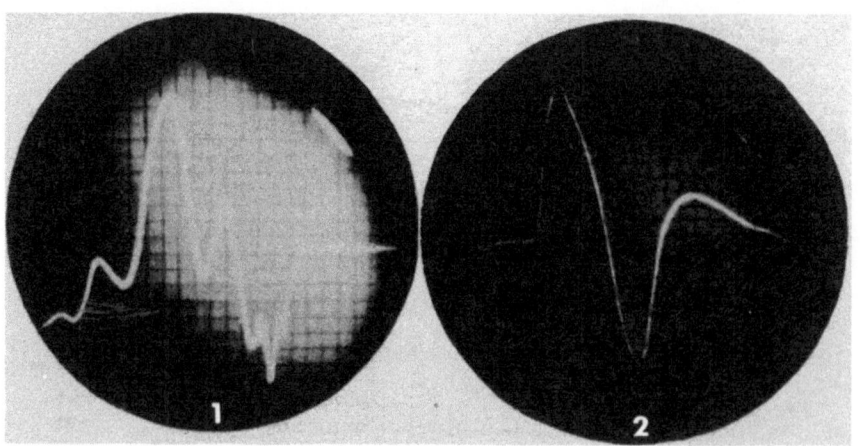

Fig. 68. Oscillograms of pressure impulses in liquid with the nozzle immersed deeply in the hole. 1) 1500 m; 2) 2500 m.

From oscillograms of pressure impulses from depths (immersion of nozzle) of 1500 and 2500 m (Fig. 68), recorded by the above-described apparatus, it may be seen that the intensity of the signals remains practically constant and noticeably exceeds the noise level [63].

In conclusion we should repeat that an acoustical method of control during explosive drilling at depth is expediently combined with the use of capsules having spacing guides for distance control of explosions.

Chapter 8

THE TECHNOLOGY OF DRILLING BORE HOLES BY EXPLOSIONS

By practice in deep drilling, the technological methods and norms have been worked out for geological conditions approximating those in oil and gas fields.

Both in construction and reinforcement of the hole as well as in measures of prevention from losses of circulating drilling fluid, water shows, and gas shows, from blowouts of gas and oil, and from slumping, drilling by explosions shares common problems with drilling by means of a bit. Therefore, our consideration of the question of technology of sinking holes by explosions will be limited to experimental investigation of the parameters defining the regime distinctively belonging to explosive drilling. These parameters are: frequency of exploding capsules at the bottom of the hole, the periodicity of feeding capsules into the hole, the rate of explosive drilling, the advance of the instrument during explosive drilling, and the consumption (and properties) of drilling fluid.

In reviewing the technological possibilities of the method, we have made an evaluation of the possibility of using explosive drilling for making a hole of small diameter.

In addition we have considered the scattering of capsules over the bottom of the hole, as this affects the effectiveness of the explosions, the diameter of the hole, and the development of cavities; we have also examined, in its widest sense, the question of bending of holes, i.e., the factors that determine the quality of a shaft drilled by explosions.

The technology of explosive drilling has been worked out for various intervals of depth during sinking of the exploitation well 855 (560-1650 m) in the Tuimazy oil field (Bashkir ASSR), in an experimental-industrial hole (650-2300 m) and the exploratory holes 1D (2235-2350 m) and 2D (2406-2807 m) in the Yablonovskii area of the Kinel'-Cherkassy oil field, and also in the industrial well 277 (700-2054 m) in the Mukhanovskii area (Kuibyshev Oblast).

The Procedure of Drilling by Explosions

The Procedure of Feeding Capsules into the Holes. The rate of explosive drilling, in meters per hour, is determined by the depth increment at the bottom of the hole for each explosion and by the hourly rate of feeding capsules into the hole, with computation of the actual number of detonated capsules.

The capsules may be fed into the hole continuously, with evenly spaced pauses, or periodically, i.e., in groups, with evenly spaced pauses between series of capsules.

In the first technique the rate of explosive drilling is determined by the frequency of feeding the capsules into the hole. In the second method the rate depends on the frequency of feeding the capsules into the well, on the number in each group, and on the length of time between groups of capsules, during which time no capsules are fed into the hole.

The procedure of feeding capsules into a hole is determined by the conditions of cleansing crushed rock fragments from the bottom and by the method used to control the distance between projectile nozzle and hole bottom.

A disturbance to the relationship among effectiveness of the explosions, the pause between explosions, and the intensity of the flushing process may lead to accumulation of sludge on the bottom, to a decrease in effectiveness of the explosions, and to the development of cavities on the walls of the hole. Here the concept of intensity of flushing involves the consumption of drilling fluid, the properties of the fluid, and the rate of fluid discharge from the projectile nozzle. It is possible to attain complete removal of sludge from the bottom and to achieve maximum effectiveness of the explosions, at a constant intensity of flushing, by a proper increase in the length of the pause between explosions, when capsules are fed continuously into the hole, or by a decrease in the number of capsules per group and an increase in the intervening pause between groups, when capsules are fed into the hole in periodic groups.

When using a conical nozzle with capsules having no spacing guides, only the periodic system of feeding is possible, since in this method the control and restoration of the initial distance between nozzle and bottom are effected periodically, in the pauses between the supply of groups of capsules, by touching the nozzle to the bottom of the hole.

A supported projectile nozzle, being in contact with the bottom at the moment of explosion, can be used with periodic or with continuous feeding of capsules into the hole.

The hourly rate of supplying capsules to the hole may be computed by the following formulae:

$$N = \frac{3600\,n}{n\,\Delta t + \Delta T}, \tag{21}$$

where N is the number of capsules fed into the hole per hour, n is the number of capsules in a group, Δt is the length of time between capsules within a group, in seconds, and ΔT is the interval of time between groups, in seconds, or

$$N = \frac{3600\,n}{T}, \tag{22}$$

where $T = n\Delta t + \Delta T$, the duration of the cycle for one group of capsules to pass from the equipping-feeding apparatus on the surface down to the bottom of the hole.

In fixing n when a conical nozzle is employed, it is necessary to consider the distance between nozzle and bottom allowable in light of possible scattering of the capsules on the bottom. Consequently, the number of capsules in a group will depend on the effectiveness of the explosions and may be designated in advance from experimental data on sinking holes through various geologic formations. When capsules are scattered, the initial distance between nozzle and bottom (about 200 mm) may be increased as the hole deepens, approximately to 400-450 mm, without substantially lowering the effectiveness.

When drilling holes with a conical projectile nozzle, the number of capsules in a group, n, is generally taken as 10, with an average effectiveness of explosions of 10 mm and above (but no greater than 20 mm); and n is increased to 20, when the effectiveness of explosions is less than 10 mm.

When supported projectile nozzles are used, the capsules may be fed into the hole continuously at a definite frequency, or large groups may be supplied with the number of capsules in each group determined by technological considerations: removal of sludge, rotation of the instrument (freely suspended), and sluggish adjustment of the instrument. In practice, the number of capsules per group ranges from 30 to 120 when a supported nozzle is used.

Other factors restricting the frequency of explosions are:

The transmission of detonations through the liquid from the explosion of an active charge on the bottom of the hole to the nearest passive charge moving through the column of drill pipe;

Strength of the capsule envelope, which is subjected to the effect of shock waves traveling through the liquid in the pipe;

The minimum time necessary for mixing the initial components and forming a charge of liquid explosive;

Derangement of the intervals between capsules on their way from the equipping-feeding apparatus to the bottom of the hole.

TABLE 34. The Relation of Frequency of Explosions on the Effectiveness of the Explosions

Hole	Horizon, brief description of rock, and depth interval	Type and consumption of drilling fluid, liters/sec	No. of capsules per group	Pause, sec between explosions in group	Pause, sec between groups of capsules	Lowering of bottom per explosion, mm	Projectile nozzle	Hourly rate of exploding capsules on bottom of hole
855	Myachkovo; limestones (cherty) and dolomites; 559–583 m	water; 45–50	1	—	30	15.0	Conical	120
			2	5	30	15.0	The same	180
			4	5	30	13.6	" "	288
	Podol'sk; cherty limestones; 614–640 m	The same	10	5	30	12.0	" "	450
Experimental industrial	Upper Carboniferous; limestones of intermediate hardness; porous; 756–1005 m	Clay mud 50–60	Continuous supply of capsules	5	—	14.8	Supported	720
				10	—	17.0	The same	360
	Myachkovo; dolomites; 1009–1207 m	The same	60	5	80	7.0	Conical	570
			60	10	80	8.7	The same	318
	Podol'sk; dolomites and limestones; 1210–1435 m	"	10	5	80	8.7	" "	280
			10	10	80	12.8	" "	200
	Kashira and Bashkiria; limestones; 1467–1538 m	"	10	5	80	9.2	" "	280
2D	Famennian and Frasnian; cherty limestones with layers of dense clays, marl; 2406–2771 m	Clay mud	20	10	120	4.4	Conical	220
			60	10	120	4.0		300

Most of the indicated factors have been considered already in relevant parts of this book; we therefore limit ourselves here merely to a brief mention of them.

The investigation of transmission of detonations through the liquid has shown that, in accordance with the technical scheme worked out for the adopted composition of liquid explosive, explosive drilling may be carried out with a frequency on the order of 20 explosions per minute (see Table 30).

The dynamic strength of the capsule envelope of the design adopted, in view of the repeated effects of shock waves traveling through the liquid in the pipe, is sufficient to preserve the components of the liquid explosive or to preserve unitary charges at distances at which they are safe from the transmission of detonations between them.

The time required to form a unitary charge of liquid explosive mixture from individually nonexplosive oxidizing agents and fuels amounts to 1.5-2.0 sec, and, consequently, this also is in agreement with the allowable frequency of exploding capsules on the bottom of the hole in light of the transmission of detonations between explosive charges.

The frequency of explosions is determined by the time interval between explosions of capsules within a group. The hourly rate of explosions of capsules on the bottom of a hole is defined by the frequency of explosions, the number of capsules in a group, the number of groups, and the time interval between groups of capsules.

The relation of the indicated factors to the effectiveness of explosions was tested in relatively homogeneous rocks, with drilling fluids of similar properties and used at similar rates; the results permit us to make the following conclusions (Table 34).

1. In agreement with the discovery that there is some correspondence between effectiveness of explosions (in holes with relatively small variations in diameter), frequency of explosions, and consumption of drilling fluid, a further increase in the frequency of exploding capsules on the bottom of the hole, such as doubling the frequency, led to a marked decrease in effectiveness (from 12.8 to 8.2 mm). This reduction is explained by poorer cleaning of crushed rock from the bottom because of the shorter time allowed for flushing between explosions when the frequency is increased.

2. The hourly rate of detonating charges without changing the frequency of explosions or the flushing procedure may be varied within wide limits with no substantial influence on the effectiveness of the explosions.

For example, according to experimental data with the greatest number of explosions (more than 13,000 in each experiment) in comparatively homogeneous rocks of the Myachkovo and Podol'sk horizons, an increase in the hourly rate of explosions of more than twice (from 280 to 570) led to a decrease in effectiveness per explosion of less than 15% (from 8.2 to 7.0 mm).

Consequently, an increase in the rate of explosive drilling, up to a certain value of some optimum frequency of explosions, is possible by increasing the hourly rate of detonating capsules (by increasing the number of capsules in each group and by decreasing the interval between groups) even despite the slight decrease in effectiveness of the explosions.

We should make mention here of the modifying effect, on the frequency of feeding capsules into a hole, produced by derangement of the intervals between capsules during their travel to the bottom of the hole.

Observations have shown that the pauses between feeding capsules into a hole may not coincide with the pauses between explosions on the bottom of the hole. As the capsules move in the current of drilling mud along the surface path and into the column of drill pipe, the intervals between separate pairs of capsules change because of imperfections in form, inadequate smoothness of the capsule-bearing conduit, turbulence in the flow of drilling mud, or breakage (in rare cases) of parts of the capsule. A multitude of data show that most deviations from the normal length of pause at depths down to 2800 m (more than 90% of them) do not exceed 1.5 sec. However, at depths below 2000 m, the disturbance of intervals may amount to as much as ±3 or ±4 sec.

After consideration of the frequency of explosions, and from data on the disturbances to intervals between capsules during experimental drilling, it was decided to make the pause 5 sec between explosions for holes down to a depth of 1500 m (12 explosions per minute), and 10 sec for depths between 1500 and 3000 m (6 explosions per minute).

The allowable rate of detonating capsules in holes down to 1000 m deep is 720 explosions per hour, and below 2000 m, 360 explosions per hour.

Procedure of Advancing the Instrument during Explosive Drilling. Capsules are fed into the well by the equipping-feeding apparatus (Chapter 7), and the instrument is advanced (according to the deepening of the hole by the explosions) by an electrical differential automatic device [64].

The following gives on outline of the process utilizing group feeding of the capsules into the hole and periodic advance of the instrument (with a conical nozzle) in the interval between groups of explosions.

Initially, i.e., before the first explosion, the instrument is raised approximately 200 mm above the bottom of the hole. As the hole is deepened by the explosions this spacing between bottom and nozzle is allowed to increase to about 400-500 mm (depending on scattering of the capsules).

The number of capsules in a group is determined by the effectiveness of the explosions and by the final distance between nozzle and bottom, which also depends on the scattering of the capsules.

The distance between capsules in the pipe within each group is uniform and depends on the frequency of supply from the surface, the rate of flow of the drilling fluid, and on the sinking rate of the capsules themselves.

A pause is made between groups of capsules, during which time no capsules are fed into the column of pipe. The pause may be of various lengths; the interval chosen is assigned by a programed relay. In the indicated example the pause is made 48 sec long. Thus, when the rate of flow of the drilling mud is 4 m/sec in the pipe (5 $9/16$-in. pipe, consumption of fluid of 50 liters/sec) and the frequency of feeding capsules and of explosion of capsules, 12 per minute, with 10 capsules in each group, it is easy to visualize the movement of the capsules along the column of pipe. The capsules move in groups, the distance between capsules in each group being 5 sec $\times$ 4 m/sec = 20 m; each group of 10 capsules extends for 180 m along the column of pipe.

The distance between the last capsule in one group and the first capsule in the following group, when the length of the interval between groups is made 48 sec, is 192 m, and the number of capsules in 1000 m of pipe in the hole for this particular procedural pattern is $\frac{1000}{180 + 192} \times 10 = 27$.

After the last capsule in a group is detonated, there begins a pause (48 sec), during which time the first capsule of the following group is passing through the segment of pipe separating it from the bottom of the hole. During this pause the column of drill pipe is automatically lowered till the conical nozzle touches the bottom, and when a certain axial load is attained, assigned by a programed pressure relay (generally 2-5 tons), the nozzle is again raised to the initial distance (200 mm) above the bottom. Thus, after each group of capsules is ejected, the column of drill pipe is made to rest part of its weight on the bottom of the hole, advancing, in this process, the initial distance of 200 mm between nozzle and hole and some supplementary distance corresponding to the increment of depth produced by the explosions of the given group of charges.

In order to curtail the unproductive time (the time spent in touching the bottom with the nozzle and then restoring the initial distance between nozzle and bottom), the number of capsules per group may properly be increased if, for the given effectiveness of explosions, this does not lead to an inadmissible increase in distance between nozzle and bottom and to excessive scattering of the capsules.

The regime of feeding the capsules and of advancing the instrument is set beforehand by proper programed relays. One of these, according to an assigned time graph, switches the motor of the equipping-feeding apparatus on and off; another operates the motor of the electrical differential automatic device for advancing the instrument. This latter motor acts to raise the instrument, and, in agreement with the same time graph through the chain drive and the transmission shaft of the drill hoist, automatically sets the position of the column of pipe for the proper spacing between nozzle and bottom of hole.

An interrupted regime of advancing the instrument and of feeding capsules into the hole, such as at a frequency of 12 per minute in groups of 10, and with a pause of 48 sec between groups, leads to a waste of approximately 50% unproductive time. It is therefore expedient to conduct explosive drilling with the maximum possible number of capsules in a group, up to continuous feeding into the hole with an assigned frequency, if, under such circumstances, the bottom of the hole may be properly cleaned and the effectiveness of the explosions is maintained as the required height.

During explosive drilling the dimensions of rock particles detached from the rock mass and the conditions of cleaning these particles from the bottom of the hole and of removing them from the hole differ substantially from the corresponding features encountered when drilling with a bit.

During explosive drilling the type of drilling fluid acquires a special significance. In mechanical drilling the duration of continuously circulating solution in the hole is determined by the stability of the bit, and does not exceed several hours.* During explosive drilling the fluid is pumped continuously, sometimes for 10 hours. If the fluid is inferior and if excessive filtration occurs, a thick filter cake is formed, and this leads to plugging, which causes jamming of the instrument in the hole and raises the pressure on the pumps.

Frequent raising and lowering operations with the drill column when drilling with a bit facilitates mechanical removal of the mud cake from the walls of the hole and prevents plugging by the mud, whereas explosive drilling is carried on with few removals (and insertions) of the instrument.

During explosive drilling, irregularities and fractures form on the walls of the hole; the filtration surface is greater than for holes drilled by bits, and, consequently, the possibility of forming mud cake is substantially enhanced.

Thus, from the viewpoint of the general position of technology in drilling holes, higher requirements are required of drilling muds during explosive drilling. In order to create normal conditions for adding to the column of drill pipe according to increasing depth of the hole, the fluid must have high static sheer strength and comparatively high specific gravity, and, according to the conditions of developing mud cake and plugs, there should be minimum water loss and the cake should be dense and thin.

Model experiments (Chapter 2) on massive rocks and experience with explosive drilling in deep holes (Chapter 3) have established the fact that sludge on the bottom of the hole shields the bottom from the effect of explosion of a comparatively small charge in the capsule, decreases the effectiveness, increases the diameter of the hole, and forms cavities. The cavities and the general increase in diameter of the hole in turn impairs the removal of large rock particles, promotes their accumulation on the bottom, and leads to decreased effectiveness of the explosions.

During experimental studies of explosive drilling in deep holes 250-350 mm in diameter, using capsules with spacing guides and with charges of the adopted explosive weighing 50 g, observations were made on the effect of fluid consumption, fluid properties, and duration of flushing on the size of sludge particles removed from the hole and on the effectiveness of the explosions. The consumption of drilling fluid (water and clay mud) ranged from 22-25 to 70 liters/sec. At the minimum consumption of fluid the failures of charges to detonate increased progressively after a comparatively small number of explosions (10-20 capsules), apparently because of extensive accumulation of crushed rock particles beneath the projectile nozzle, which prevented escape of the capsules from the conduit of the nozzle. With increased consumption of drilling fluid, to 35-40 liters/sec, the process of explosive drilling was quickly restored and for some time progressed normally. However, drilling by explosions at this consumption of drilling fluid, especially when water or inferior clay mud is used, causes rather rapid accumulation of sludge particles on the bottom, decreases the effectiveness, and increases the number of failures among the capsules. Any rather deep penetration under such circumstances is made difficult also because of occasional constriction and jamming of the drilling instrument. Under certain conditions, when flushing has been incomplete, especially when water is used as the drilling fluid, considerable difficulty arises when the drill column is lengthened by the addition of a section of pipe, since the cessation of circulation, even for a brief time, leads to rapid settling of the large rock particles that have not been flushed from the hole.

An increase in consumption from 40-50 to 65-70 liters/sec does not produce the same increase in effectiveness of explosions as the change from 22-25 to 35-40 liters/sec, but in this case, especially when using high-grade drilling mud, more large rock fragments are washed from the hole. At this consumption rate large intervals (as much as 100 m) are drilled without complications during the time of adding pipe.

The state of the hole, development of cavities, depth, and curvature of hole have been shown to have a substantial influence on the flushing of sludge from the hole during explosive drilling.

---

*We have in mind a rolling cutter bit.

Experience has demonstrated that when water is used for flushing (consumption of 50-60 liters/sec, upward velocity of fluid of 0.8-1.0 m/sec), large fragments of dense dolomite and limestone, weighing as much as 100 g and more, may be removed from vertical holes from a depth of 1000 m when cavities in the walls of the hole are not numerous.

At the same intensity of flushing with clay mud (density of 1.19 g/cm^3, viscosity of 22-25 sec, filtration of 30 cm^3 in 30 min, and static shear stress of 30-40 mg/cm^2), rather fine sludge particles weighing at most 7-10 g were washed from a depth below 2000 m, from a hole having several cavities of various lengths and having diameters ranging from 500 to 700 mm.

With somewhat better clay mud (density of 1.22 g/cm^3, viscosity of 27 sec, filtration of 20 cm^3, and static shear stress of 40 mg/cm^2) and with a longer period of flushing, particles weighing up to 20 g and more were washed from the hole from depths of more than 2500 m.

The rate of settling of sludge in clay mud depends primarily on the specific gravity and the static shear stress of the mud. When these parameters are increased respectively to 1.26 g/cm^3 and 200 mg/cm^2, larger fragments weighing as much as 40 g may be removed from a deep, cavity-lined, and partly curved hole.

It was noted previously (Chapter 6) that, apart from the particles of sludge of the indicated size removed from the hole, under unfavorable circumstances it is possible that larger rock fragments, whose removal by flushing is practically impossible, may collect on the bottom of the hole (from the hydraulic zone of explosive influence).

When this happens, the effectiveness of explosions declines because of repeated crushing of these large rock fragments by the explosions, till they become fine enough to be removed by the fluid from the hole.

Therefore, together with the recommendations made from the results of investigating hydraulic removal of large rock fragments from the hole (Chapter 6), it is necessary to consider the possibility of decreasing the maximum dimensions of the particles detached from the rock mass during underwater explosions in the hole.

Of the two likely solutions to this problem—decreasing the size of charge or making a corresponding change in the properties of the explosive (parameters of the explosion)—we shall examine only the first. At the same time small charges may be used for explosive drilling of holes with smaller diameter.

The Scattering of Explosive Charges over the Bottom of the Hole

The movement of the capsules through the pipe is speeded up by the flow of drilling fluid in the channel of the projectile nozzle; the capsules are then ejected from the nozzle and detonated on impact with the bottom of the hole.

Experiments on models and computations have shown that scattering of the ejected capsules over the surface of the bottom substantially modifies the effectiveness of explosions and the diameter of the hole.

It is possible to distinguish hydrodynamic and geometric scattering of the paths of movement of the capsules. The first is due chiefly to asymmetry in the movement of the capsules as they are ejected from the projectile nozzle and may occur when the channel of the nozzle coincides fully with the axis of the hole, whereas the second type result from lack of coincidence of nozzle and axis of hole.

The relationship between scattering of capsules over the floor and the effectiveness may be determined by the declension of the effectiveness:

$$\gamma = \frac{h_{av}}{h_0},$$

(23)

where h_{av} is the increment of depth per explosion and h_0 is the penetration produced by a single explosion.

It is obvious that the greatest value of h_{av} corresponds to the circumstance in which all points of impact of the charges coincide, i.e., when $h_{av} = h_0$ and $\gamma = 1$.

The coefficient γ may be expressed in a functional relationship in the form

$$\gamma = f\left(\frac{D_{100}}{R_{crush}}\right),$$

where D_{100} is the diameter of the smallest circle corresponding to 100% impact zone of the charges, and R_{crush} is the radius of the opening of the funnel of crushing from an explosion on the plane of the bottom of the hole.[*]

A study based on experimental data has shown that when the ratio D_{100}/R_{crush} increases from zero to unity, i.e., when all the points of impact are within the radius R_{crush}, the value of γ is near unity, and, consequently, the average penetration h_{av} is near the maximum penetration h_0. With further increase in the ratio D_{100}/R_{crush}, γ decreases noticeably and, as shown by approximate calculations, when $D_{100}/R_{crush} \approx 3.0$-$3.5$, $\gamma = 0.3$-0.25, which corresponds to a decrease in penetration to $^1/_3$-$^1/_4$ the greatest value.

According to experimental data on a concrete block, when the charges are not placed in the center of the hole, but, for instance, at four points around the circumference, the penetration in the model hole is diminished to approximately one-fourth and the diameter is approximately doubled; these results are in complete agreement with calculations obtained with the relationship in (24).

The consequence of scattering of capsules on the effectiveness of explosions has been tested by sinking different segments of holes in different rocks with conical and supported projectile nozzles having a smooth column of drill pipe; a column of pipe having a centering device was also used on a connection with a conical nozzle.

The use of a device for centering the conical nozzle in the hole when drilling soft (nonslumping) rocks or rocks of intermediate hardness interbedded with softer layers does not lead to any marked increase in effectiveness of explosions over a smooth column weighted in the lower part with thick-walled drill pipe. In experiments conducted in sections composed of the indicated rocks, the centering of the conical nozzle either led to no higher average effectiveness of explosions or this increase did not exceed 10-20%. When drilling in hard, very cherty limestones and dolomites, the centering of the conical nozzle raised the effectiveness in some experiments 50-60%.

The use of supported projectile nozzle (on a smooth drill column having a weighted lower end) gave a noticeable increase (up to 30%) in effectiveness, in a number of experiments, and led to a hole of somewhat smaller diameter (in soft, weak, and intermediate rocks) than produced by a conical nozzle on a column with no centering device. This may be explained by less scattering of the capsules because of the minimum distance between nozzle and bottom fixed by the attached props.

However, when drilling by explosions at depths greater than 2200-2500 m, the effectiveness of the explosions decreases approximately to a common value, and there is considerable decrease in the diameter of the hole whether the work is done with a conical nozzle (both with and without centering device) or with a supported nozzle.

It may be assumed that at great depths the negative effect of technological factors on effectiveness of explosions (such as inadequate cleaning of the large particles from the bottom of the hole) and the compaction of rocks noticeably offset the results of centering the nozzle.

Thus, the centering of a nozzle in the hole decreases the scattering of the capsules over the bottom, increases the effectiveness of explosions, and diminishes the diameter of the hole. The consequence of scattering on the effectiveness of explosions is clearly manifested in strong rocks, more susceptible to fracturing and spalling, and is less noticeable in soft weak rocks and such rocks interbedded with harder rocks.

Peculiarities of the Explosive Method of Drilling Holes

During explosive drilling the diameter of the projectile nozzle and of the pipe has been proved to have no direct effect on the diameter of hole formed. By using charges of different sizes and with various distributions over the bottom of the hole, it is possible to regulate the diameter of the hole within wide limits and, in particular, to drill deep oil and gas wells of minimal diameter allowable for production from the well, such as 6 inches and less. In addition, it is possible to cover certain segments of the hole, which complicate the sinking of the entire hole,[**] with a column of pipe (with minimum loss of diameter) and to pass on through to produce

---

[*] The actual form of the bottom is not planar, and because of this R_{crush} is corrected by a suitable factor.
[**] Beds with high pressure, shows of water and gas, water-absorbing horizons, segments complicated by unstable slumping rocks.

an open shaft of the required diameter in the uncomplicated part of the section. Thus, conditions favorable for drilling in the deep intervals may be created by using ordinary (unweighted) clay muds, water, or gas.

In explosive drilling it is characteristic to obtain comparatively large rock fragments, detached from the mass, chiefly from the walls of the hole in the hydraulic zone of explosive activity. The size of the spalling fragments depends on the properties of the rock being drilled, and it is possible to regulate this size by varying the size of the charge and the parameters of the explosive. Large rock fragments brought to the surface are of special value in studying the section in the hole.

The possibility of transmittting energy to the bottom of the hole with practically no loss, a factor that limits the depth of hole when using mechanical methods of drilling, permits one to use the principal advantages of explosive drilling for sinking holes to very great depths (to 7 km and more), if, under these conditions, the change in properties of the rocks because of high pressure and temperature does not lead to a sharp decline in the susceptibility of the rocks to explosions.

The first indicated peculiarity of the explosive technique compares with the modern trend in deep drilling, i.e., smaller diameter of the hole.

In moving on to consider the basic parameters characterizing explosive drilling of small-diameter holes, we should recall that the depth of the funnel, from the local brisance effect of exploding the charge to the contact with the rock, as well as the ultimate diameter of the hole, is approximately proportional to the cube root of the weight of the charge (Chapter 2). Comparative data on the effectiveness of explosions on concrete blocks, lead targets, and in actual holes, with charges of different sizes, are shown in Table 35 (average results).

In testing model charges on various concrete blocks (series I and II of the experiments) it has been shown that the actual increase in depth of the model hole per explosion amounts to from 71 to 103% of the theoretical increase expected, and the diameter attained amounts correspondingly to from 81 to 134%.

The divergence between experimental and theoretical data is apparently connected in considerable measure with the fact that when changes were made in size of charge, changes were also made in the shape of charge and the envelope confining it; the effects of these factors when working with model charges are rather considerable.

The results of tests on actual (in size and form) explosive charges on massive lead targets has shown a near correspondence between expected and actual depth of the funnel [$\sqrt[3]{(32/59)} = 0.815$ and $\sqrt[3]{(19/23)} = 0.826$, respectively].

Comparative data on drilling with different charges under approximately similar conditions make it plain that the average increment of depth per explosion is 88 to 102% of the theoretical value expected.

Thus, in evaluating the results of comparable experimental and calculated data, we may conclude that the relationship (7) in general satisfactorily describes the connection between size of explosive charge and effectiveness (penetration) per explosion. The restrictions on using the indicated relationship are, in particular, the dimensions of the charge, the diameter of which should exceed the critical value, which is the minimum necessary for normal development of detonation.

The critical diameter of a charge for suitable liquid explosive is measured in parts of a millimeters, and a conversion of the experimental results obtained during tests on a model charge of 0.46 ml to an actual charge of 33 ml shows a satisfactory correspondence of the experimental results for the large charge.

If, when using the explosive technique to drill holes with explosive charges of weight (volume) G_0, a diameter of hole of D_0 is formed with a penetration per explosion of h_0, a change to drilling a hole with a diameter of

$$D = \frac{D_0}{K} \tag{25}$$

in rocks of the same explosive susceptibility and with a charge similar in form and position of the initiating element (and similar construction of the capsule), would require a charge weighing

$$G = \frac{G_0}{K^3} ; \tag{26}$$

the penetration per explosion under these conditions would be

$$h = \frac{h_0}{K} \ .\tag{27}$$

Figure 69 furnishes a graphic representation of the relationship between size of charge and the coefficient of scale conversion K. This relationship is basic for the following analysis.

TABLE 35. Comparative Effectiveness of Explosions of Charges of Various Sizes on Concrete Blocks, Lead Targets, and in Holes

Specimen no.	Size of charge G, ml	$\frac{G}{G_0}$*	$\sqrt[3]{\frac{G}{G_0}}$	Actual deepening per explosion h_a, mm	Theoretical deepening per explosion $h_t = h_0{}^* \times \times \sqrt[3]{\frac{G}{G_0}}$, mm	$\frac{h_a}{h_t}100$, %	Actual diameter of hole D_a, mm	Theoretical diameter of hole D_t, mm	$\frac{D_a}{D_t}100$, %
Concrete blocks under a 6-m layer of water (series I of experiments)
1	2.4	1.000	1.000	7.26	7.26	100.0	87	87.0	100
2	1.2	0.500	0.793	4.11	5.75	71.5	56	69.0	81
3	0.46	0.192	0.578	3.22	4.19	77.0	54	50.3	107
Concrete blocks under a 6-m layer of water (series II of experiments)
4	1.2	1.000	1.000	3.75	3.75	100.0	57	57.0	100
5	0.46	0.383	0.726	2.76	2.74	100.5	49	42.5	115
6	0.23	0.191	0.576	2.32	2.16	103.0	44	32.8	134
Lead targets under water
7	59	1.000	1.000	23	23.0	100.0			
8	32	0.542	0.815	19	18.7	101.5			
Experimental-industrial hole (736-770 m)
9	59	1.000	1.000	11.27	11.27	100.0			
10	32	0.542	0.815	8.05	9.15	88.0			
Experimental-industrial hole (1210-1232 m)
11	32	1.000	1.000	8.07	8.07	100.0			
12	18	0.563	0.825	5.89	6.65	90.0			
Hole 1D (2348-2359 m)
13	59	1.000	1.000	6.05	6.05	100.0			
14	32	0.542	0.815	5.00	4.90	102.0			

*The subscript 0 indicates parameters relating to the initial charge.

According to the data of explosive drilling with charges of $G_0 = 32$ ml, the average diameter of a hole ranges from 300 to 350 mm. In order to discover the technique of drilling a hole with, for example, half the diameter, i.e., D = 150-175 mm, it is necessary to have the following initial data, illustrated graphically in Fig. 69:

$$\frac{D_0}{D} = K = 2; \quad G = \frac{G_0}{K^3} = \frac{32}{2^3} = 4 \ \text{ml} \cdot \quad h = \frac{h_0}{K} = \frac{h_0}{2} \ \text{mm}; \quad \frac{L_0}{L} = 2.$$

Coarseness of the Sludge. When the form of the charge and the conditions of explosion are constant and the same rock is involved, the total volume of shattered material W is proportional to the size of charge. Consequently,

$$\frac{W_0}{W} = \frac{G_0}{G} = K^3,\tag{28}$$

where W_0 is the volume of shattered rock when drilling with charges of G_0. At $K = 2$

$$W = \frac{W_0}{8}.$$

Large fragments of rock spall off generally from the walls of the hole at distances somewhat removed from the focus of the explosion at (or near) the bottom of the hole (hydraulic zone of explosive action); the spalling is due to shock waves traveling through the liquid during an explosion.

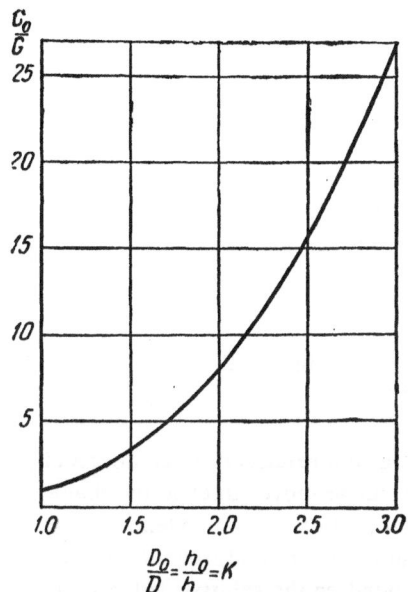

Fig. 69. Relationship between relative size of charge (G_0/G) and the coefficient of scale conversion K.

The largest rock fragments, only partly flushed out or not removed at all, fall to the bottom, screen the bottom from the effect of the explosion, and decrease the effectiveness of the explosion.

The size of the fragments detached from the mass of rock depends on the depth of fractures formed in the rock and on the dimensions of the surface developed about the bottom of the hole, and this latter, in turn, is determined by the diameter of the hole.

In order to determine the maximum size of sludge particles when changing the procedure to drilling holes of small diameter by decreasing the explosive charges, we must make the following assumptions.

1. The depth δ of the fractures (see Fig. 70), which defines the thickness of the particles, is proportional to the shattering impulse of pressure of the shock waves I_{sh} traveling through the liquid.

2. The average surface area S of the large rock fragments A reflected toward the axis of the hole is proportional to the area of surface in the near-bottom zone of the hole (or the cross-sectional area of the hole) from which the fragment is detached.

In view of the fact that if the relative distance of the indicated rock fragment A from the focus of the explosions remains constant, $R/a =$ constant (a is the radius of the charge employed), on the basis of our assumptions it is possible to write for the similar elements — the sludge fragments A_0 and A (see Fig. 79)— in large and small holes

$$\frac{\delta_0}{\delta} = \frac{I_{0sh}}{I_{sh}} = \sqrt[3]{\frac{G_0}{G}} = \frac{D_0}{D} = K; \tag{29}$$

for surfaces of similar elements it will be valid to use the equation

$$\frac{S_0}{S} = \left(\frac{D_0}{D}\right)^2 = K^2. \tag{30}$$

Consequently, the ratio of weights (or volumes) of the large sludge particles in large and small holes is

$$\frac{\delta_0 S_0}{\delta S} = K^3. \tag{31}$$

Thus, when changing to a procedure of drilling a hole with half the diameter, the weight of the large sludge particles will be but one-eighth the weight of the particles in the larger hole.

Flushing. If, in the smaller hole, the same rate of flow of the drilling fluid is maintained (at the bottom and in the annular space between pipe and wall), the intensity of flushing should increase markedly. When the diameter of the hole is reduced to one-half, the consumption of drilling fluid may be diminished from 60 to 15 liters/sec.

When the diameter of the hole is reduced by a factor of K, the amount of sludge material per explosion is decreased by a factor of K^3 and the area of the bottom by a factor of K^2; i.e., the amount of sludge material falling on a unit area of the bottom surface is reduced by a factor of K when the rate of flow of drilling fluid is maintained.

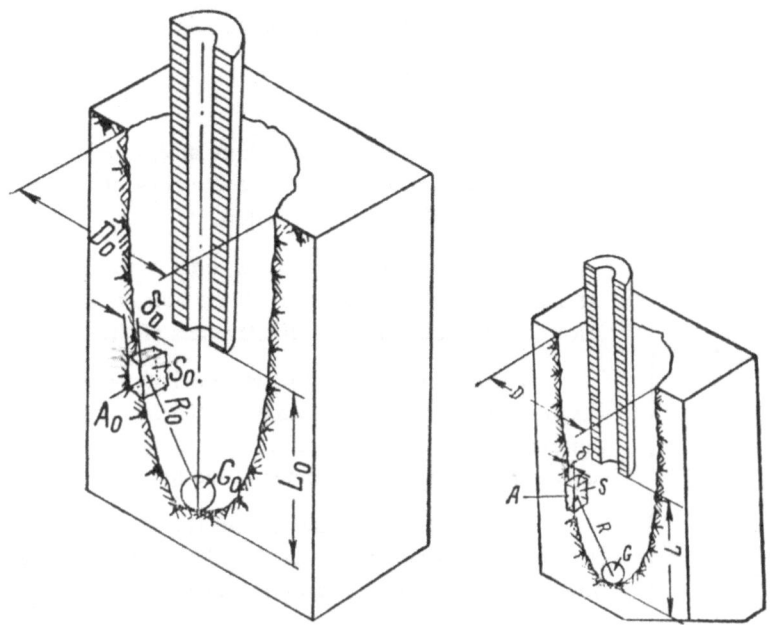

Fig. 70. Schematic drawing of some similar elements during explosive drilling of holes with explosive charges of G_0 and of $G = G_0/2^3$.

During cleaning operations on the bottom, when the sludge particles acquire a relatively large acceleration, the forces overcome are chiefly inertial forces. Subsequently, when the sludge moves through the space between pipe and wall, the rise of particles is hindered primarily by the weight (gravitational tendency to sink). In both situations the forces are proportional to the volume of the particles and are decreased, for the smaller hole, by a factor of K^3. The active forces transporting the sludge particles depend on the relative velocity v with streamlined flow of the particles, on the form and size of the particles, on the density γ of the particles, and on the viscosity of the fluid. According to Shushchenko [59] spherical particles with a diameter d and moving in a viscous liquid have the following relation to the force P that sets the particles in motion, for various values of Reynolds' number: at Re $\leq$ 5

$$P = 3\pi v \mu \left(1 + \frac{3}{16} \operatorname{Re}\right) d; \tag{32}$$

at $5 < \text{Re} \leq 1000$

$$P = 3\pi v \mu (1 + 0.15 \operatorname{Re}^{0.687}) d; \tag{33}$$

at Re > 1000

$$P = c \frac{\pi d^2 \gamma v^2}{4g}. \tag{34}$$

Here μ is the dynamic coefficient of viscosity; for water at a temperature of 20°C, $\mu = 10^{-4}$ kg $\cdot$ sec/m^2. According to Rettinger, when Re $\geq$ 1000, c = const = 0.26 for a sphere (with dimensions measured in centimeters, and time in seconds).

An examination of these formulae shows that when K = 2, for all practical values of the relative rate of flow, the forces setting the particles in motion are diminished by no more than the factor K^2, i.e., by an amount approximately proportional to the surface area of the particle.

Thus, when decreasing the diameter of a hole to one-half, the force for moving and subsequently lifting particles of similar size, considered per unit volume, doubles. This should lead to a relative increase, in holes of small diameter, of the limiting weight of large particles removed from the hole and should also lead to faster cleaning of the bottom and removal of the smaller particles from the hole.

Frequency of Explosions. In order to preserve the rate of explosive drilling (such that $h = h_0/K = h_0/2$) when the diameter of the hole is reduced to one-half ($D_0/D = K = 2$), it is necessary to cut the penetration per explosion in half to compensate for the doubled frequency of explosions of the small charges (4 ml). This is possible, keeping in view the possible transmission of detonations, because, when the linear dimensions of such charges are changed by a factor of n, or, correspondingly, the weight of the charge is changed by a factor of n^3, the critical distance between charges changes by a factor greater than n.

Stability of the Projectile Nozzle. When the linear dimensions of a charge are cut in half and the relative distance between a given point and the center of a spherical charge is maintained, $R/a = $ constant, the maximum pressure at the front of the shock wave remains unaltered. However, the length of time the shock wave acts and, consequently, the magnitude of the pressure impulse change proportionally with the dimensions of the charge; i.e., in this example they are reduced to one-half.

When the diameter of the hole is reduced to one-half, a corresponding decrease in the diameter of the instrument does not lead to a reduction in the strength of the conical nozzle; consequently, when small charges are denoted, the nozzle is actually operating under more favorable conditions. Therefore, it is possible to consider the nozzle to have greater stability when we make calculations for accidental disturbance to the controlled distance between nozzle and bottom of hole, when the nozzle may be nearer the explosion center than generally acceptable.

Indices for Explosive Drilling of Holes with Small Diameter. K times as many small charges are required as large charges to maintain the same penetration rate. Therefore, λ (the cost ratio of unit penetration by small and large capsules) will be

$$\lambda = \frac{aK}{A},$$

where A and a represent total cost of large and small capsules, respectively.

The coefficient λ describes the economy of drilling holes of small diameter with capsules of the given design and explosive composition. When $K = 2$ the volume of the charge is reduced to one-eighth, and the requirements for material of which the capsules are made are diminished by about the same figure.

On the basis of an over-all analysis of the changes in expenditures when using small charges (4 ml) during explosive drilling, it has been established that for almost all examined capsule designs (the linear dimensions of which were reduced by one-half) the cost of unit depth of sinking a hole 150 mm in diameter as compared to the comparable cost for a hole 300 mm in diameter gives a value of λ of 0.41-0.83.

Starting from the general principles of this relationship, we may find confirmation that a change in the diameters of a hole and of the drill pipe to one-half leads to the following changes:

The consumption of drilling mud, cement, metal drill pipe and casing, and material for the capsules is cut by a factor of K^2.

The decrease in weight of the column of drill pipe by this factor of K^2 (at a value of $K = 2$ and more) permits a substantial lessening of the weight of the drilling apparatus as a whole.

As a consequence of the decrease in volume and weight of the capsule (when $K = 2$), of the drill pipe, and in the consumption of mud, cement, and other materials necessary for the explosive drilling, storage operations are simplified and are approximately one-quarter as expensive, and transportation operations are also greatly lessened.

Because of a considerable reduction in power of the pumping system and of the drive on the hoist, the expenditure of energy is decreased.

Other Procedures of Explosive Drilling of Holes. Until now we have considered the plan of explosive drilling of holes by explosive charges of various sizes directed toward the center of the bottom of the hole.

With centrally placed explosions, the diameter of the hole formed is determined by the size of charge of a given explosive. For drilling oil and gas wells of the required diameters, rather large charges are used with this technique. Moreover, when the rocks are shattered by centrally placed detonations, spalling is meagerly developed in the rocks about the zone of the local explosive effect.

In another scheme we may consider forced scattering of the explosive charges over the bottom of the hole, the size of the charges being very small in comparison with the area of the bottom of the hole [65]. In this procedure, for most of the explosive charges, the walls of the hole exert an insignificant influence on the shattering of the rock.

During the explosion of a very small charge (0.4 ml of liquid explosive), the volume of shattered rock derived from the explosive charge is approximately twice that from a charge detonated in the center of the bottom, and the penetration per explosion is increased to 1.5 times. The increase in volumetric productivity and effectiveness of explosions in this technique is explained by more intense spalling.

The rate of explosive drilling, when this plan is employed, according to preliminary experimental data, should be greater than the rate with centrally placed explosions of larger explosive charges, because of the increased frequency of explosions, greater efficiency of the explosions, better cleaning of sludge from the bottom, and greater stability of the projectile nozzle.

Another advantage of this technique of drilling is the possibility of regulating the diameter of the hole by changing the dimensions of the charge or the radius of their scattering over the bottom of the hole (while maintaining the same size of charge).

Preference should be given to charges of liquid explosive without envelopes when drilling by this technique (using the bottom apparatus previously discussed).

It is assumed also that a detachable conical nozzle will be used, which may be removed and exchanged without difficult and prolonged hoisting and lowering operations with the drill pipe.

The possibility of rapid removal of the projectile nozzle, apart from the general increased rate of sinking deep holes, permits periodic cleaning of the accumulated rock fragments from the hole by reversing the circulation of drilling fluid.

Moreover, when considerable thicknesses of clay are encountered, through which penetration by explosives is ineffective, it is advisable to combine explosive drilling, using the detachable nozzle, with rotary drilling, using a detachable bit. In this way it may be possible to achieve continuous sinking of extensive intervals in deep holes.

Curvature of Holes during Explosive Drilling. An analysis of the results of sinking long intervals by explosives in holes down to 2800 m deep has shown that, when operating with a supported projectile nozzle, bending of the shaft occurs, in some holes reaching inclinations of 17-20°.

When working under similar conditions with a conical projectile nozzle, the segment of hole produced proved to be practically vertical (a curvature within 1°).

Observations have shown also that when a conical nozzle is used in a curved hole, the curvature may be eliminated, and the axis of the hole becomes approximately vertical within a very short segment when the hole is sunk farther. This may be explained by the fact that the column of pipe with the conical nozzle hangs above the bottom during the process of explosive drilling, whereas the supported nozzle may become inclined by its contact with the bottom, against which it is pressed by the weight of the drill column.

Chapter 9

RESULTS AND POTENTIALS OF EXPLOSIVE DRILLING

Principal Results

Experiments in explosive drilling, begun in an experimental hole, were soon transferred to oil districts in the eastern part of the country, the geology of which is distinguished chiefly by strong rocks.

Under such conditions of industrial operation, experimental explosive drilling was conducted in depth intervals ranging from 540 to 2800 m.

The possibility of drilling deep holes without a bit was first established in Tuimazy (Bashkiria), where a hole was sunken by explosions through cherty limestones in the depth interval 540-700 m (hole 208). At the same time it was found that the technical factors were improved, and only a very inconsiderable segment of the shaft produced by explosive drilling had large cavities (see Fig. 37).

The first commercially valuable results were obtained in the same region, where a hole was drilled 31 m by uninterrupted explosive operation through cherty limestones and dolomites, with an average rate of about 4 m/hr (hole 855). In this hole more than 100 m was drilled by explosions with much less cavity development in the shaft (see Fig. 38). An experiment was conducted in the same hole, at a depth of over 1640 m, to use explosives to open up a productive (oil-bearing) sandstone (see Fig. 39).

Future work, as the method was developed, led to increased depth of drilling, improvement in the shaft of the hole, and increase in one of the basic indices: the total penetration for each removal of the instrument.

Nearly 1000 m were drilled by explosions in an experimental-industrial hole in the Yablonovskii area (Kuibyshev Oblast) at various intervals of depth down to 2300 m. In the Mukhanovo field (Kuibyshev Oblast), 720 m were drilled by explosions in hole 277 in the depth interval 697-2054 m. The greatest interval drilled without a bit (about 300 m) was drilled in the same hole, and a penetration of 113 m was obtained for a single projectile nozzle.

In the experimental-industrial hole and in hole 277, at depths down to 1500-2000 m in hard rocks, penetrations of 100 m for each removal of the instrument were repeatedly obtained, and the over-all rate of drilling (considering these removals) exceeded the rate of drilling with bits.

The greatest penetration rate with explosive drilling, equal to 16 m/hr, was obtained in limestones of intermediate hardness at a depth of 900 m.

During experimental operations in the Yablonovskii area, in very strong Famennian and Frasnian cherty limestones in the exploratory hole 2D at a depth of 2250-2745 m, penetrations of as much as 20 m were obtained for each lowering of the instrument into the hole, and the over-all rate of drilling, with the raising and lowering operations included, amounted to 0.6 m/hr; this compares with an average penetration per bit, under similar conditions, of about 3 m, and an over-all drilling rate, including the raising and lowering operations, of 0.15 m/hr.

The greatest penetration for a single projectile nozzle at depths greater than 2400 m was 41 m, and the maximum rate of explosive drilling under these conditions did not exceed 2 m/hr.

The segments of deep holes drilled by explosive means, by using a conical nozzle, were practically vertical.

The Effect of Hole Depth on Effectiveness of Drilling

Leaving for the moment the question concerning the effects of rock properties on explosive drilling, let us examine the over-all picture of the changes in the process with depth, i.e., the penetration (effectiveness) and volume of shattered rock per explosion (the volumetric productivity of an explosion).

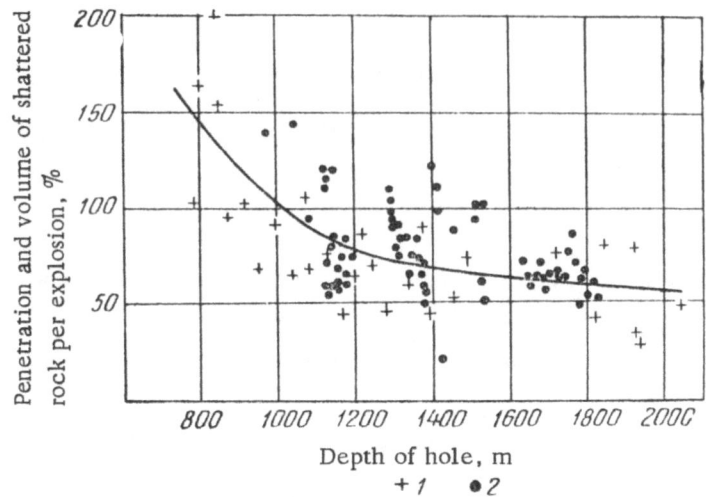

Fig. 71. Penetration and volume of shattered rock per explosion in relation to depth of hole. 1) Penetration; 2) volume of shattered rock.

Figure 71 shows graphically the changes in effectiveness of explosions with depth (according to drilling data from hole 277). From this graph it may be clearly seen that the effectiveness of explosions decreases with increasing depth of hole. A similar relationship is general for explosive drilling under various geologic conditions.

The probable causes of the decreased effectiveness of explosions with increasing depth of hole are associated with the previously discussed factors effecting changes in rock properties (Chapter 3) and the shattering of rocks by explosions (Chapters 2 and 3), and also with technological factors, of which cleaning crushed rock from the bottom of the hole has been shown to have the greatest consequence on the effectiveness of explosions (Chapters 6 and 8).

The influence of the various factors on the effectiveness and on the volumetric productivity of an explosion may be traced in a supplementary fashion on the graph of Fig. 72. Each point on the graph corresponds to a depth of hole, measured in hundreds of meters. An analysis of this graph confirms the tendency toward decreased volumetric productivity of an explosion with depth of hole. At the same time, it may be noticed that the points referring to the greatest depths (this group of points is enclosed in a box in Fig. 72) lie somewhat below the general curve relating penetration per explosion to volume of shattered rock per explosion for most of the experiments. For the indicated group of points, in moving to the right, it is characteristic that, despite a slight rise in volumetric productivity of the explosions, the penetration per explosion diminishes. This points to a predominance of technological factors, and, above all, it reflects the difficulty of cleaning the bottom of the hole in the worsening conditions at greater depths.

In studying the graph of Fig. 72 we may note that, though there is no functional connection, there is a well-defined correlation between the considered values. The general distribution of points indicates that a change in penetration per explosion parallels a change in volume of shattered rock per explosion. Different rocks (in geologic structure, lithic properties, and mode of occurrence) possess substantially different susceptibilities to explosions.

Thus, in all the examined instances, an over-all decrease in effectiveness and in volume of shattered rock is observed with increasing depth of hole. Other conditions being equal, in going from a depth of about 1000 m to a depth of 2000 m and more, the penetration per explosion, for the explosive charges adopted, decreases to one-half or one-third.

The decrease in penetration and volume of shattered rock per explosion may be explained not only by worsening of the explosive susceptibility of the rocks (according to their compaction by hydrostatic and rock pressure), but also by the ever-possible technological complications associated with deep drilling.

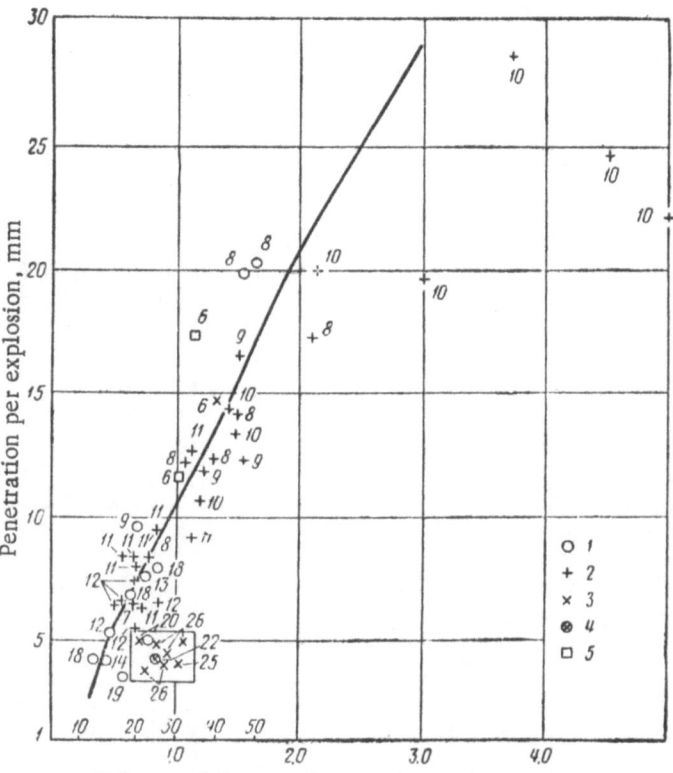

Fig. 72. Relationship between penetration and volume of shattered rock per explosion. 1) Hole 277; 2) experimental-industrial hole; 3) hole 2D; 4) hole 1D; 5) hole 855.

The Effect of Rock Properties on Effectiveness of Drilling

Changes in rock properties prove to have a strong influence on the effectiveness of explosions. This is confirmed by the results from all the holes, and may be illustrated by the experimental-industrial hole and hole 277, the sections of which, in those segments of the holes drilled by explosions, are characterized by frequent alternations of different beds (especially in hole 277).

Let us examine one of the segments drilled by explosions in hole 277 (2039-2054 m) made up of cherty limestones (the Tula "plate"). This formation is underlain by a layer of clay about 10 m thick. Explosive drilling in this plate was carried on with an average effectiveness per explosion of 5 mm; toward the end of drilling through the plate the effectiveness increased to 7.5 mm.*

---

* A slight increase in effectiveness has also been observed in model experiments at the boundary where strong hard rocks grade into softer rocks.

On passing from a hard rock into clay, the drilling process exhibited a sharp drop in effectiveness. A similar phenomenon was observed in hole 2D during attempts to drill the Kyn clay (the Middle Devonian infra-Domanik formation) at a depth of 2830 m.

In some of the segments of the experimental-industrial hole drilled by explosions, a change in effectiveness, caused by a change in the properties of the rock drilled through, appeared rather abruptly; this was obviously due to individual fluctuations in effectiveness because of various technological factors.

Figure 73 shows a composite diagram for several segments of this type. The diagram has the superimposed curves of apparent resistivity (taken from three sondes), self-potential, caliper logs, a histogram of changes in effectiveness, and the lithic section. One may trace on this diagram the separate abrupt changes in effectiveness at the beds of rock distinguished by logging data.

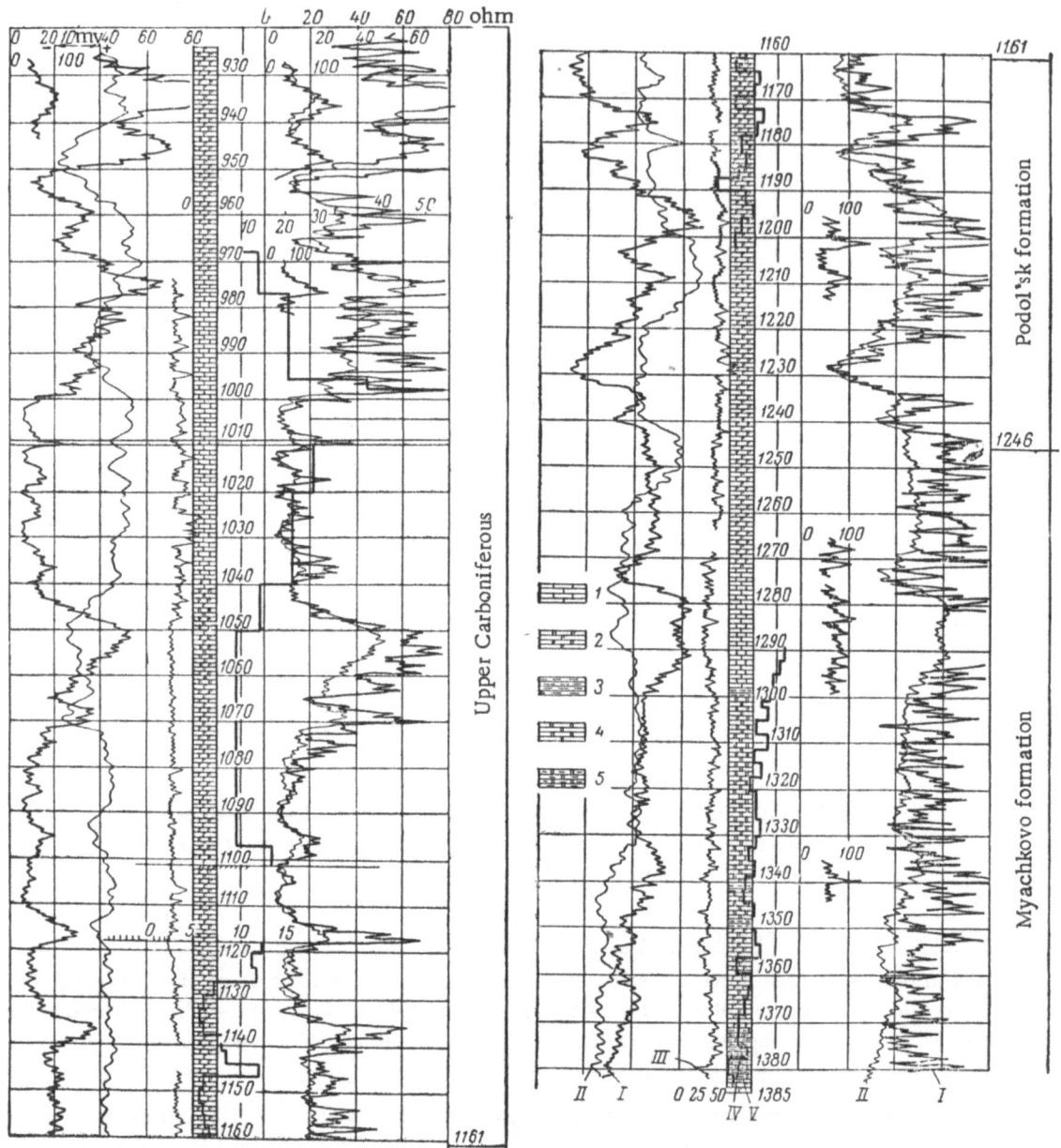

Fig. 73. Composite diagram of a segment of the experimental-industrial hole (Yablonovskii area) drilled by explosions. I) Apparent resistivity log; II) self-potential log; III) caliper log; IV) histogram of changes in effectiveness of explosions; V) descriptive lithology; 1) organic-fragmental limestone; 2) dolomitized limestone; 3) clay; 4) dolomite; 5) dolomitized argillaceous limestone.

The sharp increase in effectiveness in the interval 995-1004 m was associated with a bed of cavernous limestone.

The logs for hole 277 (Mukhanovo area) are less complete than those for the experimental-industrial hole (Yablonovskii area). Furthermore, the section at the Mukhanovo area is characterized by a large a number of comparatively thin heterogeneous beds. Because of these it becomes difficult to associate segments of the section passed through with different values of effectiveness, to identify these values with particular beds. Comparisons are possible, however, for some segments (Table 36).

As may be seen from the composite diagram (Fig. 73) and from Table 36, the boundaries of the intervals drilled through by explosions in which a marked change in effectiveness was noticed to coincide rather closely with the formational boundaries distinguished on the logs.

Above, as an extreme case, we noted a sharp change in the effectiveness of explosions when drilling clays and extremely porous cavernous limestones. As a characteristic example we may point further to beds of gypsum

TABLE 36. A Comparison of Beds Distinguished on the Log and of Boundaries of Intervals Drilled through with Varying Effectiveness of Explosions, from Hole 277

Boundaries of beds distinguished on the log, m		Boundaries of segments distinguished by average effectiveness of explosions, m		Average penetration per explosion	
top	bottom	top	bottom	mm	%
792.0	797.0	794.0	798.0	18.0	200.0
797.0	799.0	798.0	803.0	27.0	300.0
822.0	827.0	821.5	825.4	10.7	119.0
831.0	837.0	829.5	836.0	28.0	312.0
837.0	841.0	836.0	841.0	19.0	211.0
841.0	850.0	841.0	849.0	32.0	356.0
850.0	853.0	849.0	853.0	16.0	178.0
853.0	869.0	853.0	868.0	9.0	100.0
1117.5	1119.0	1117.5	1119.5	13.7	254.0
1119.0	1123.0	1119.5	1121.5	5.8	107.0
1123.0	1127.0	1121.5	1126.5	13.8	256.0
1127.0	1130.0	1126.5	1128.5	8.5	157.0
1130.0	1133.0	1128.5	1133.5	14.3	265.0
1145.0	1149.0	1146.0	1148.0	24.3	451.0
1149.0	1152.0	1148.0	1153.0	5.4	100.0
1154.0	1158.0	1153.0	1156.5	10.4	192.0
1158.0	1161.5	1156.5	1162.4	7.0	129.0
1222.0	1226.0	1222.0	1226.7	6.7	167.0
1226.0	1229.5	1226.7	1229.7	14.8	370.0
1229.5	1231.0	1229.7	1231.7	4.7	117.0
1231.5	1234.0	1231.7	1233.6	6.1	152.0
1235.0	1238.0	1233.6	1237.6	11.4	285.0
1238.0	1239.0	1237.6	1241.6	8.7	218.0
1247.0	1249.0	1241.6	1249.5	4.0	100.0
1251.0	1254.0	1249.5	1253.5	10.9	273.0

*For a more precise evaluation of the effect of rock properties, three intervals are distinguished in the table according to depth of the hole, each of which has a minimum value of effectiveness, taken as 100%.

and anhydrite. For these rocks, and also for sections of carbonate rocks with high gypsum content, the penetration rate and the volume of shattered rock per explosion are characteristically rather low. A shaft of smaller diameter is formed in gypsum and anhydrite.

A high rate of penetration per explosions was observed during the drilling of very hard rocks such as cherty limestones and dolomites. A good rate was also obtained during explosive drilling of homogeneous beds of such rocks as the Myachkovo-Podol'sk series in Bashkiria (hole 855).

Analysis of Over-all Penetration Rate

The drilling of deep holes with rolling cutter bits by means of motors lowered to the bottom of the hole is characterized by rather high mechanical rate and poor penetration for a single lowering of the instrument into the hole.

The drilling of corresponding rocks with a diamond bit and by the explosive method is distinguished by high penetration per cutting interval with a comparatively low mechanical and explosive rate [66].

It is well known that the over-all penetration rate is one of the basic criteria for evaluating the effectiveness of drilling; it may be determined by the equation:

$$v_p = \frac{L}{T_{r,1} + T_{dril} + T_{inc}} , \qquad (35)$$

where L is the penetration for a single operation of the instrument in the hole (penetration per cutting interval) in meters, $T_{r,1}$ is the time required for raising and lowering the instrument in hours, T_{dril} is the length of time of actual drilling in hours, and T_{inc} is the time spent in adding to the instrument or drill column in hours.

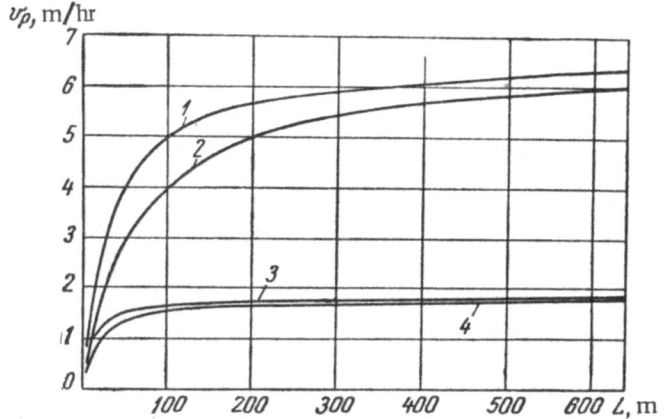

Fig. 74. Relationship between over-all penetration rate v_p and the penetration per cutting interval L for different mechanical (explosive) rates of drilling v (2 and 10 m/hr), at depths H of 1250 and 2500 m. 1) v = 10 m/hr; H = 1250 m; 2) v = 10 m/hr, H = 2500 m; 3) v = 2 m/hr, H = 1250 m; 4) v = 2 m/hr, H = 2500 m.

By substituting

$$T_{dril} = \frac{L}{v} \text{ and } T_{inc} = t_{inc} \frac{L}{l} ,$$

where v is the average mechanical (explosive) rate of drilling per cutting interval in meters per hour, t_{inc} is the time for adding a single pipe to the column in hours, and l is the average length of pipe in meters, we obtain

$$v_p = \frac{v}{1 + v \left(\dfrac{T_{r,1}}{L} + \dfrac{t_{inc}}{l} \right)} . \qquad (36)$$

Figure 74 furnishes a graphical representation of the relationship $v_p = f(L)$ in the depth intervals H of 1250 m and 2500 m for v = 2 and 10 m/hr. Figure 75 shows graphically the relationship $v_p = f_1(v)$ for several values of L. In constructing the graphs a norm was adopted for time spent in raising and lowering the column and in adding to the column (or the instrument).

124

A study of the graphical relationship $v_p = f(L)$ shows that when the value of v is low (2 m/hr) there is a rapid rise in the over-all penetration rate with an increase in L up to 60-80 m, a smaller increase when L is increased to 100-120 m, and a very slight increase when L is greater than 150 m. Thus, when drilling a hole in the interval 1250-2500 m and with L = 150 m, the average value of v_p is 1.66 m/hr, i.e., 83% of v, and when L = 300 m, v_p increases 4% in all (to 83%). Even when the indicated interval (1250-2500 m) drilled with a single penetration at v = 2 m/hr, v_p increases, according to computation, only 3% as compared with the v_p corresponding to L = 300 m. At higher values of v (10 m/hr), v_p increases more uniformly.

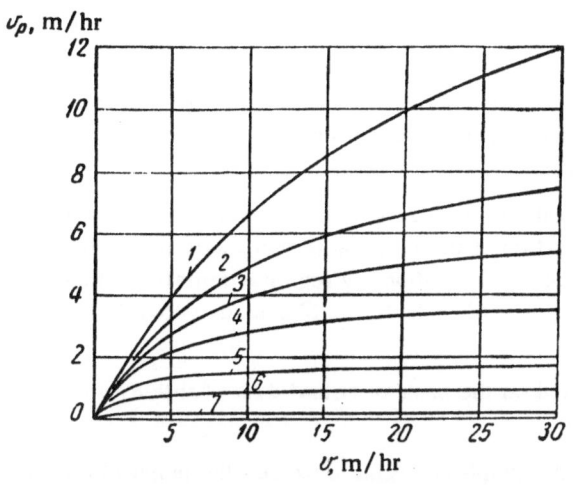

Fig. 75. Relationship between over-all penetration rate v_p and the mechanical (explosive) rate of drilling v for different penetrations per cutting interval L at the depths H. 1) H = ∞, L = ∞ (drilling without raising the instrument); 2) H = 1250 m, L = 100 m; 3) H = 1250 m, L = 50 m; H = 2500 m, L = 100 m; 4) H = 2500 m, L = 50 m; 5) H = 2500 m, L = 20 m; 6) H = 2500 m, L = 10 m; 7) H = 2500 m, L = 2 m.

From the graphical relationship $v_p = f_1(v)$ it may be seen that at a depth of 2500 m in the hole and with L ranging from 2 to 20 m, an increase in v above 3 and 5 m/hr has practically no effect on increasing v_p.

A substantial increase in v_p with an increase in v occurs in the indicated depth of hole (2500 m) only when there is a considerable increase in L (50 m and more). For comparatively shallow holes (such as 1250 m), a noticeable increase in v_p with increase in v begins at lower values of L.

From an examination of the graphical relationships for holes 2500 m deep it follows that an increase in v of more than 3-5 m/hr (up to 30 m/hr and more) produces no increase in v_p at values of L up to 20-30m.

However, at low values of v (2 m/hr) an increase in L greater than 60-80 m in the depth interval 1250-2500 m (up to a penetration of the indicated interval in a single cutting interval) also leads to no noticeable increase in v_p.

Thus, a substantial increase in the effectiveness of drilling deep intervals (above 2000-2500 m) with rolling cutter bits is possible if a penetration per cutting interval can be made greater than 50 m; with diamond bits and non-bit (explosive) methods the effectiveness is noticeably increased by increasing the mechanical and explosive rate of drilling at these intervals to 10 m/hr and more. This conclusion may be used also for evaluating the effectiveness of other methods of drilling deep holes.

Potentials of the Explosive Methods of Drilling Deep Holes

The results of experimental explosive drilling in deep holes have shown, above all, the necessity of preventing a decrease in effectiveness of drilling at increasing depth, of offsetting, if not completely at least in great measure, the negative effect increased depth has on the effectiveness of explosions.

In order to preserve the effectiveness of explosions during drilling * it is necessary to remove the causes leading to the formation or the accumulation or large rock fragments on the floor of the hole, fragments that are incapable of being flushed out and thus shield the bottom from the effect of the explosive charge. Improvement in the quality of the drilling mud (higher thixotropic tendency, static shear stress, and specific gravity, in addition to decreased water loss) and a slight increase in intensity of flushing have made it possible to remove large rock particles (several tens of grams) from deep holes.

During operations with high-grade drilling mud and when making penetrations of 100 m per cutting interval at a depth of about 2000 m, drilling was effected with practically stable effectiveness of explosions.

----

* We have in mind rocks that possess comparatively similar susceptibilities to explosions.

In order to prevent the formation of excessively large rock fragments, provisions are also made to use small charges for explosive drilling in sinking holes of reduced and small diameters (Chapter 8).

To increase the effectiveness of explosions at great depths, it is necessary to continue investigations on charges (shape, directive properties of the explosion) and on the explosives (density, detonation velocity, specific energy of an explosion). In addition, it is proposed to increase the efficiency of the explosions by proper distribution of small charges on the bottom, which will increase the spalling effects of the explosions.

In evaluating the potential of the explosive method of drilling holes, its characteristics are compared with actually obtained and expected characteristics of drilling with a turbodrill (with three-cone rolling cutter bits), with high-speed diamond bits operating on submerged motors (2000-2200 rpm), and with submerged motors in a procedure of drilling without raising the pipe [67].

In Tables 37 and 38 and on the graphs of Fig. 76* are shown the changes in over-all penetration rate v_p in relation to the depth of the hole H, computed on the basis of actual data and of anticipated potential characteristics that may be foreseen for the different methods of drilling.

In our analysis the characteristics for drilling with diamond bits were adopted from data obtained under the geological conditions found at the Lac field (France). A comparison of these characteristics with those obtained by drilling with three-cone rolling cutter bits and with explosions in the Kuibyshev region is tentative, since the results of preliminary tests with diamond bits during the drilling of hard rocks in the Mukhanovo area do not support the apparent superiority of this technique.

From a comparison of over-all penetration rates, computed on the basis of actual data (Table 37 and graphs 1, 3, and 5 in Fig. 76), it follows:

1) That the over-all penetration rate by explosive drilling (graph 1) begins to exceed the penetration rate of three-cone rolling cutter bits below 1660 m (graph 5), and at a depth of 3000 m the rate is more than 2.5 times as great.

2) That turbodrilling with diamond bits (graph 3) down to a depth of 2000 m has a markedly lower penetration rate than explosive drilling (graph 1); at 2400 m the two become equal, and at a depth of 3000 m the diamond instrument has a penetration rate 1.43 times that of explosive drilling.

A study of the anticipated results (Table 38 and graphs 2, 4, and 6 on Fig. 76) indicates the following:

1) The assumed over-all rate of explosive drilling (graph 2) down to a depth of 2000 m is greater than the anticipated rate with high-speed diamond bits using a submerged motor (graph 4). The rates become equal at a depth of 2400 m, and below this, to a depth of 3000 m, the high-speed diamond bits is 1.45 times as fast as explosive drilling, a relationship similar to that found in comparing results obtained from actual data (graphs 1 and 3);

2) The assumed over-all rate of drilling without raising the pipe, by means of a detachable submerged motor and bit (graph 6) begins at a depth of 1740 m to fall below the rate of explosive drilling (graph 2), and at depths exceeding 2080 m it falls below the rate of a high-speed diamond bit (graph 4); the over-all penetration rate of these two techniques at 3000 m are, respectively, 1.7 and 2.5 times the rate of drilling with detachable gear without raising the pipe.

The over-all penetration rates with a diamond bit with detachable gear and without raising the pipe, and by

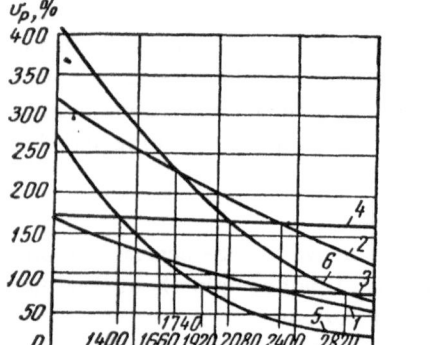

Fig. 76. Relationship of over-all penetration rate v_p and depth of hole H for various methods of drilling. 1) Explosive drilling (actual data); 2) explosive drilling (anticipated characteristics); 3) turbodrilling with diamond bits (actual data); 4) drilling with diamond bits and a high-speed submerged motor (anticipated characteristics); 5) turbodrilling with three-cone rolling cutter bits (actual data); 6) drilling without raising the drill pipe, by means of a detachable submerged motor and three-cone rolling cutter bits (assumed characteristics).

_____
*Prepared by V. D. Kruglov.

TABLE 37. Changes in Over-all Penetration Rate v_p with Depth H of Hole (actual data)

Curve on Fig. 76	Method of drilling	Over-all penetration rate									
		$H = 1000$ m		$H = 1500$ m		$H = 2000$ m		$H = 2500$ m		$H = 3000$ m	
		m/hr	%	m/hr	%	m/hr	%	m/hr	%	m/hr	%
1	Explosive	3.3	100.0	2.6	100.0	2.00	100.0	1.5	100.0	1.05	100.0
5	Turbodrilling*	5.5	167.0	3.0	115.0	1.50	75.0	0.7	47.0	0.40	38.0
3	Turbodrilling with diamond bit**	1.8	54.5	1.7	65.5	1.65	82.5	1.6	107.0	1.50	143.0

*Hole 424 (Mukhanovo area, Kuibyshev Oblast), drilled by three-cone rolling cutter bits and showing the best characteristics in the region.

**Data taken from turbodrilling under the geologic conditions at the Lac oil field (France).

TABLE 38. Changes in Over-all Penetration Rate v_p with Depth H of Hole (anticipated results)

Curve on Fig. 76	Method of drilling	Over-all penetration rate									
		$H = 1000$ m		$H = 1500$ m		$H = 2000$ m		$H = 2500$ m		$H = 3000$ m	
		m/hr	%	m/hr	%	m/hr	%	m/hr	%	m/hr	%
2	Explosive	6.4	100.0	5.2	100.0	4.1	100.0	3.1	100.0	2.2	100.0
4	Turbodrilling or electrodrilling with a high-speed diamond bit (2000–2200 rpm)	3.5	54.5	3.4	65.5	3.3	80.5	3.3	106.5	3.2	145.5
6	Without raising the pipe, by a detachable submerged motor (turbodrill, electrodrill) and a three-cone rolling cutter bit	8.3	130.0	5.7	110.0	3.6	88.0	2.1	68.0	1.3	59.0

explosions are rather high compared to the mechanical (and explosive) rates of drilling (i.e., rates when actually cutting).

Figure 77 shows graphs (prepared by A. I. Gol'binder) of explosive penetration v and the over-all rate v_p in relation to the depth of hole H (hole 277, Mukhanovo area).

On graph 1 each horizontal mark corresponds to a penetration distance for a single cutting interval. The average rate is referred to the middle of the segment of hole for each cutting interval, and is connected to the next value by a straight line. According to the drilling data on this hole, the average over-all rate is 73% of the drilling rate by explosions (that is, while actually performing drilling operations);[*] a maximum value of

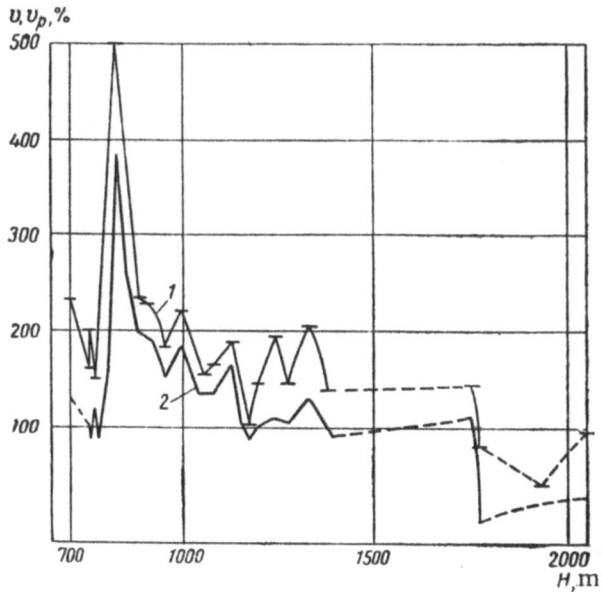

Fig. 77. 1) Relationship between rate of explosive drilling v and depth of hole H; 2) the same for over-all penetration rate v_p.

83% was achieved at a depth of 1000 m, and a minimum value of 51% was obtained at 2000 m.

For comparison we may note that, under similar conditions and at the same depth interval, a rush drilling operation with three-cone rolling cutter bits gave a relative value for over-all penetration rate at 800 m of 48% of the mechanical (actual cutting); at a depth of 1000 m the value is 35% and near 2000 m it ranges from 7 to 25%.[*]

Table 39 shows the expense of drilling one meter, both by explosive method and by using bits.

As may be seen from the cited data, the expense per meter of a hole drilled by the explosive method increases much more slowly below 2000 m than the expense of drilling with three-cone rolling cutter bits. Explosive drilling at a depth of 3000 m is less expensive by a factor of 2.6.

The assumed expense per meter of drilling with a diamond instrument at a depth of 3000 m in nonabrasive and weak rocks and in rocks of intermediate hardness is less than the expense of drilling with three-cone rolling cutter bits by a factor of approximately 5.7.

Thus, drilling with a diamond bit in combination with a high-speed submerged motor (turbodrill, electrodrill) and with non-bit drilling (explosive in this case), each for the proper rocks, represents the most promising means of increasing effectiveness of penetrating deep intervals in oil and gas wells.

---

[*] This takes into account operations with very small penetration, when, for various reasons, drilling was interrupted and the gear was removed from the hole.

[**] Hole 424 (Mukhanovo area).

TABLE 39. Expense per Meter of Drilling Holes at Various Depths*

Method of drilling	Drilling depth					
	1000 m		2000 m		3000 m	
	rubles	%	rubles	%	rubles	%
Turbodrilling (from actual data in hole 424 at Mukhanovo)	116	100.0	371	100	1675	100.0
Turbodrilling or electrodrilling with high-speed (2000-2200 rpm) diamond bits** (anticipated results)	169	146.0	233	63	296	17.7
Explosive*** (anticipated results)	254	219.0	404	109	636	38.0

* The computations take into account the expense in drilling due to type of bit and method of drilling. No consideration was given to construction not directly related to drilling the hole, or to fixed expenses not depending on the method of drilling.

** For these computations the technical indices are adopted from the conditions obtained at the Lac field (France), and the value of the diamond bit is taken as 20,000 rubles (preliminary tests with diamond bits at the Mukhanovo field did not support the apparent superiority over three-cone rolling cutter bits).

*** The capsules are evaluated at 2 rubles each.

Still, it has already been noted that the existing and newly developed methods of drilling deep holes do not furnish a universal solution; they do not satisfy in full measure the many conditions of drilling to be found in the various oil and gas fields.

Drilling with diamond bits, after one has solved some of the problems associated with obtaining the required quantity of large diamonds, apparently has but a limited range of application, as is true with explosive drilling. For example, preliminary data indicate that it is inadvisable to use diamonds, because of their tendency to chip out, when drilling abrasive, siliceous, hard, and igneous rocks. Under such conditions explosive drilling is more suitable.

Consequently, rolling cutter bits, diamond bits, and explosives are not mutually exclusive means of drilling, and a correct combination of them for proper geologic conditions, kinds of rock, and depth of drilling may lead to more effective drilling of deep oil and gas wells.

The use of the explosive method makes it possible to drill deep oil and gas wells of minimum diameter suitable for actual production (Chapter 8). In using this method with small explosive charges, the stability of the bottom instrument (projectile nozzle) is considerably increased. This fact is all the more important in view of the extreme difficulty facing the manufacture of a high-strength rolling cutter bit with a diameter less than 190 mm and of an immersible motor having high energy parameters and a diameter less than 170 mm.

Apart from solving the task of deep drilling applicable to the geologic conditions in oil and gas fields, explosive drilling (both with capsules and with uncontained explosive charges) may be used for sinking exploratory holes at various depths and for drilling in hard crystalline rocks the so-called "blasting" holes in open workings on ore deposits.

When there is no inflow of formational water, when there is little show of water or absorption of the drilling fluid, explosive drilling may be carried out with air flushing of the bottom or with flushing by natural gas obtained from neighboring wells. In this process, for any particular explosive charges, the formation of large rock fragments incapable of removal from the hole is excluded, the total volume of shattered rock is reduced (to $\frac{1}{5}-\frac{1}{10}$), and the effectiveness of explosive drilling may be improved, since the penetration per explosion should not decrease as the depth of hole becomes greater. Furthermore, with due consideration to the transmission of detonations between charges, the frequency of explosions and, correspondingly, the rate of explosive drilling may be increased fivefold and more.

Explosive drilling should also find application in the future development of sinking holes at depths greater than 5-7 km.

LITERATURE CITED

1. N. S. Timofeev. "Development of the petroleum industry in the Kuibyshev Oblast in 1959-1965." Neftyanoe khoz., No. 1, 1959.
2. M. T. Gusman. "The use of diamond bits in turbodrilling." Neftyanoe khoz., No. 1, 1959,
3. A. P. Ostrovskii and Ya. M. Kershenbaum. "A hydraulic giant arrangement for drilling holes and doing other mine work." Author's certificate No. 129581 of Dec. 9, 1944. Byulleten' izobretenii Komiteta po delam izobretenii i otkrytii pri Sovete Ministrov SSSR, No. 12, 1960, Moscow.
4. A. P. Ostrovskii, A. A. Pavlichenko, V. M. Slavskii, N. G. Taurok, and F. F. Voskresenskii. "A hydraulic arrangement for drilling holes and doing other mine work." Author's certificate No. 73481 of March 14, 1947 Byulleten' izobretenii Komiteta po delam izobretenii i otkrytii pri Sovete Ministrov SSSR, No. 3, 1949.
5. M. Kornfel'd. Elasticity and Strength of Liquids. GITTL [State Publishing House for Technical and Theoretical Literature], 1951.
6. V. O. Mal'chenok and O. M. Sumarkov. "The prospects of using oscillating processes in drilling." Trudy Vsesoyuznogo nauchno-issledovatel'skogo instituta metodiki i tekhniki razvedki (VITR), Collection 1. Gosteoptekhizdat [State Scientific and Technical Publishing House of the Petroleum and Mineral Fuel Industry], 1958.
7. A. P. Pinsker, D. E. Tagamlik, and M. K. Kogan. Investigations on Mine Construction. The Possibility of Using Ultrasonics in Shattering Rocks. All-Union Scientific-Research Institute for Organization and Mechanization of Mine Construction, 1956.
8. L. Bergman. Ultrasonics and Its Use in Science and Technology [Russian translation]. IL [Foreign Literature Publishing House], 1956.
9. L. A. Yutkin. The Electrohydraulic Effect. Mashgiz [State Scientific Publishing House of Literature on Machinery], 1955.
10. N. I. Titkov, M. A. Varzanov, I. I. Slezinger, O. P. Petrova, and G. I. Borisov. "Drilling by means of electical discharges in a liquid." Neftyanoe khoz., No. 10, 1957.
11. P. A. Kulle and P. V. Ponomarev. Trudy Vsesoyuznogo nauchno-issledovatel'skogo instituta metodiki i tekhniki razvedki (VITR), Collection 1. Gostoptekhizdat [State Scientific and Technical Publishing House of the Petroleum and Mineral Fuel Industry], 1958.
12. L. Leb. Fundamental Processes of Electrical Discharges in Gases. Gostekhteoretizdat [State Publishing House for Technical and Theoretical Literature], 1950.
13. L. A. Sena. Collision of Electrons and Ions with Atoms of Gas. Gostekhteoretizdat [State Publishing House for Technical and Theoretical Literature], 1948.
14. Frungel. Optik, Vol. 3, 1948.
15. M. A. Sadovskii. "The local effect of an explosion." Trudy Seismicheskogo instituta AN SSSR, No. 11, 1941, Moscow.
16. A. I. Gol'binder and A. P. Ostrovskii. "Experimental investigation on the explosive drilling of holes." Razvedka nedr, No. 6, 1959.
17. A. P. Ostrovskii. "Formation of holes by blasting." Novosti neftyanoi tekhniki. GosINTI, No. 10, 1958.
18. E. E. Stetyukha. Determination of the Most Important Factors Affecting the Mechanical Rate of Drilling at Depths below 3000 m. Dissertation. I. M. Gubkin Moscow Petroleum Institute, 1955.
19. A. P. Ostrovskii, A. A. Pavlichenko, V. M. Slavskii, A. I. Gol'binder, and N. G. Grigoryan. "A method of drilling holes and doing other mine work by explosions, and the equipment used in employing this method." Author's certificate No. 121390 of August 3, 1956. Byulleten' izobretenii Komiteta po delam izobretenii i otkrytii pri Sovete Ministrov SSSR, No. 15, 1959.

20. A. P. Ostrovskii, A. A. Pavlichenko, V. I. Slavskii, A. I. Gol'binder, and N. G. Grigoryan. " An arrangement for drilling holes by blasting." Author's certificate No. 122107 of Sept. 30, 1947. Byulleten' izobretenii Komiteta po delam izobretenii i otkrytii pri Sovete Ministrov SSSR, No. 17, 1959.

21. L. A. Shreiner. The Physical Basis of Rock Mechanics. Gostoptekhizdat [State Scientific and Technical Publishing House of the Petroleum and Mineral-Fuel Industry], 1950.

22. U. B. Brooks, L. U. Henderson, and F. U. Shell. New Methods of Drilling [Russian translation]. Fourth International Petroleum Congress. Gostoptekhizdat [State Scientific and Technical Publishing House of the Petroleum and Mineral-Fuel Industry], 1958, Vol. 3.

23. A. V. Brichkin, T. E. Zhakupov, and P. Ch. Chulakov. "Theoretical and computed grounds for the design of a thermodrill." Gornyi zhurnal, No. 4, 1957.

24. I. P. Goldaev, E. P. Polevichek, N. N. Popov, and A. P. Pershin. "Thermal drilling of massive rock by jet reaction." Byulleten' nauchno-tekhnicheskoi informatsii. GosINTI, No. 14, 1958.

25. A. V. Bruchkin, P. Ch. Chulakov, and A. N. Genbach. "Conditions for intensive drilling of rock by the thermal method." Vestnik Akademii nauk Kazakhskoi SSR, 1957.

26. A. Engel' and M. Shtenbek. The Physics and Technique of Electrical Discharge in Gases. ONTI [United Scientific and Technical Publishing Houses], 1936, Vol. 1.

27. G. Cann and A. Ducati. Energy Content and Ionization Level in an Argon Gas Jet Heated by a High Intensity Arc. Journal of Fluid Mechanics, Vol. 4, No. 1, 1958.

28. G. Sutton. Journal of the Aeronautical Sciences, Vol. 25, No. 1, 1958.

29. N. A. Kaptsov. Electrical Phenomena in Gases and in a Vacuum. Gostekhteoretizdat [State Publishing House for Technical and Theoretical Literature], 1950.

30. Dzhianini (Gianini). Atomnaya tekhnika za rubezhom (Atom Techniques in Foreign Countries), No. 2, 1958.

31. H. Maecker. Z. Physik, Vol. 129, 1951.

32. R. W. Larenz. Z. Physik, Vol. 129, 1951.

33. M. A. Gintsburg. "Shattering rocks by high-frequency electromagnetic fields." Izvestiya Akademii nauk SSSR, No. 10, 1957.

34. All-Union Coal Institute. Development of Electrophysical Methods for Crushing Coal and Rock. Ugletekhizdat [State Scientific and Technical Publishing House of Literature on the Coal Industry], 1957.

35. V. S. Kravchenko. "The search for new methods of shattering strong rocks." Gornyi zhurnal, No. 1, 1957.

36. "Circular waveguides for distant transmission of centimeter- and millimeter-sized waves." Cables et Transmission, Vol. 2, No. 4, 1957.

37. A. Sivers and N. Suslov. Fundamentals of Radar. Sovetskoe radio [Soviet Radio], 1957.

38. M. S. Neiman. A Course in Radio Transmitting Equipment. Sovetskoe radio [Soviet Radio], 1958, Part 2.

39. G. I. Pokrovskii and I. S. Fedorov. The Effect of Impact and Explosions in Strained Media. Gosudarstvennoe izdatel'stvo literatury po stroitel'nym materialam [State Publishing House for Literature on Structural Materials], 1957.

40. A. I. Gol'binder, E. B. Kagan, and A. P. Ostrovskii. "The mechanism of drilling holes by blasting." Neftyannoe khoz., No. 7, 1958.

41. L. D. Landau and K. P. Stanyukovich. "Studies of detonation in condensed explosive materials.'' Doklady Akademii nauk SSSR, Vol. 46, 1945.

42. Yu. B. Zel'dovich and K. P. Stanyukovich. "Reflections of plane detonation waves." Doklady Akademii nauk SSSR, Vol. 55, 1947.

43. K. P. Stanyukovich. "The emission of detonation products from oblique detonation waves." Doklady Akademii nauk SSSR, Vol. 55, 1947.

44. K. P. Stanyukovich. Irregular Movements of Massive Media. Gostoptekhizdat [State Scientific and Technical Publishing House of the Petroleum and Mineral-Fuel Industry],1956.

45. R. Cole. Underwater Rxplosions [Russian translation]. IL [Foreign Literature Publishing House], 1952.

46. Journal of Petroleum Technology, No. 11, 1955.

47. Oil and Gas Journal, No. 10, 1955.

48. V. S. Fedorov. Bits for Drilling Oil Wells. Azgostoptekhizdat [Azerbaidzhan State Scientific and Technical Publishing House of the Petroleum and Mineral-Fuel Industry], 1941.

49. O. E. Vlasov. Fundamentals of the Theory of Explosive Action. Kuibyshev Military-Engineering Academy, 1957.

50. F. A. Baum, K. P. Stanyukovich, and B. I. Shakhter. The Physics of Explosions. Fizmatgiz [State Publishing House for Physics and Mathematics], 1959.

51. Byurlo. Mediate Detonations. Publishing House of the Artillery Academy of the Workers' and Peasants' Red Army, Leningrad, 1933.

52. K. J. Eichelberger and M. Sultanoff. Proceedings of the Royal Society, Vol. 296, 1958.

53. W. R. Marlow and J. C. Skidmore. Proceedings of the Royal Society, Vol. 246, 1958.

54. H. Lawton and J. C. Skidmore. Discussions Faraday Society, Vol. 22, 1956.

55. C. H. Winning. Proceedings of the Royal Society, Vol. 246, 1958.

56. C. H. Winning. Comptes rendus du XXVII Congres international de chimie industrielle. Brussels, 1954.

57. Bowden and Joffe. The Ignition and Development of Explosions in Solid and Liquid Materials [Russian translation]. IL [Foreign Literature Publishing House], 1955.

58. E. B. Kagan and A. P. Ostrovskii. "Investigations on the removal of large rock particles from holes." Neftyanoe khoz., No. 4, 1959.

59. R. I. Shishchenko. Hydraulic Clay Muds. Azgostekhizdat [Azerbaidzhan State Publishing House of Technical Literature], 1951.

60. A. P. Ostrovskii. "An explosive method of drilling holes and doing other mine work." Author's certificate No. 117234 of June 5, 1956. Byulleten' izobretenii Komiteta po delam izobretenii i otkrytii pri Sovete Ministrov SSSR, No. 1, Moscow, 1959.

61. A. P. Ostrovskii, V. K. Bogomolov, G. V. Vladimirov, A. I. Gol'binder, V. V. Grachev, V. D. Kruglov, Zh. I. Shnapir, and V. L. Shukhman. A capsule for explosive drilling of holes and for other mine work." Author's certificate No. 126839 of June 5, 1956. Byulleten' izobretenii Komiteta po delam izobretenii i otkrytii pri Sovete Ministrov SSSR, No. 6. Moscow, 1960.

62. A. P. Ostrovskii, V. K. Bogomolov, G. V. Vladimirov, F. F. Voskresenskii, V. V. Grachev, and V. D. Kruglov. "A feeding-equipping apparatus." Author's certificate No. 126840 of June 5, 1956. Byulleten' izobretenii Komiteta po delam izobretenii i otkrytii pri Sovete Ministrov SSSR, No. 6. Moscow, 1960.

63. Ya. V. Rubinovich and V. P. Varlamov. "Hydraulic communication channels to hole bottom for turbodrilling." Neftyanoe khoz., No. 3, 1958.

64. A. P. Ostrovskii. "A device for automatically advancing the drill assembly." Author's certificate No. 53624 of Nov. 13, 1937. Byulleten' izobretenii, No. 8, 1938.

65. V. K. Botomolov, V. N. Grinblat, and E. B. Kagan. "A capsule explosive drill." Author's certificate No. 118790 of May 27, 1958. Byulleten' izobretenii Komiteta po delam izobretenii i otkrytii pri Sovete Ministrov SSSR, No. 7, 1959.

66. A. P. Ostrovskii and V. D. Kruglov. "An analysis of over-all drilling rate." Trudy VNIIBT, 1960.

67. A. P. Ostrovskii. "The potentials of the new methods of drilling holes." Neftyanoe khoz., No. 9, 1959.